Grade K

Reveal MATH®

Student Practice Book

McGraw Hill

mheducation.com/prek-12

Send all inquiries to:
McGraw Hill
8787 Orion Place
Columbus, OH 43240

ISBN: 978-0-07-693701-1
MHID: 0-07-693701-1

Printed in the United States of America.

8 9 10 BRR 26 25 24 23 A

Grade K
Table of Contents

Unit 2
Numbers to 5

Lessons

Unit 3
Numbers to 10

Lessons

Unit 4
Sort, Classify, and Count Objects
Lessons

Unit 5
2-Dimensional Shapes
Lessons

Unit 6
Understand Addition
Lessons

Unit 7
Understand Subtraction
Lessons

Unit 8
Addition and Subtraction Strategies
Lessons

Unit 9
Numbers 11 to 15

Lessons

Unit 10
Numbers 16 to 19

Lessons

Unit 11
3-Dimensional Shapes

Lessons

Unit 12
Count to 100

Lessons

Unit 13
Analyze, Compare, and Compose Shapes

Lessons

Unit 14
Compare Measurable Attributes

Lessons

Additional Practice

Name _____

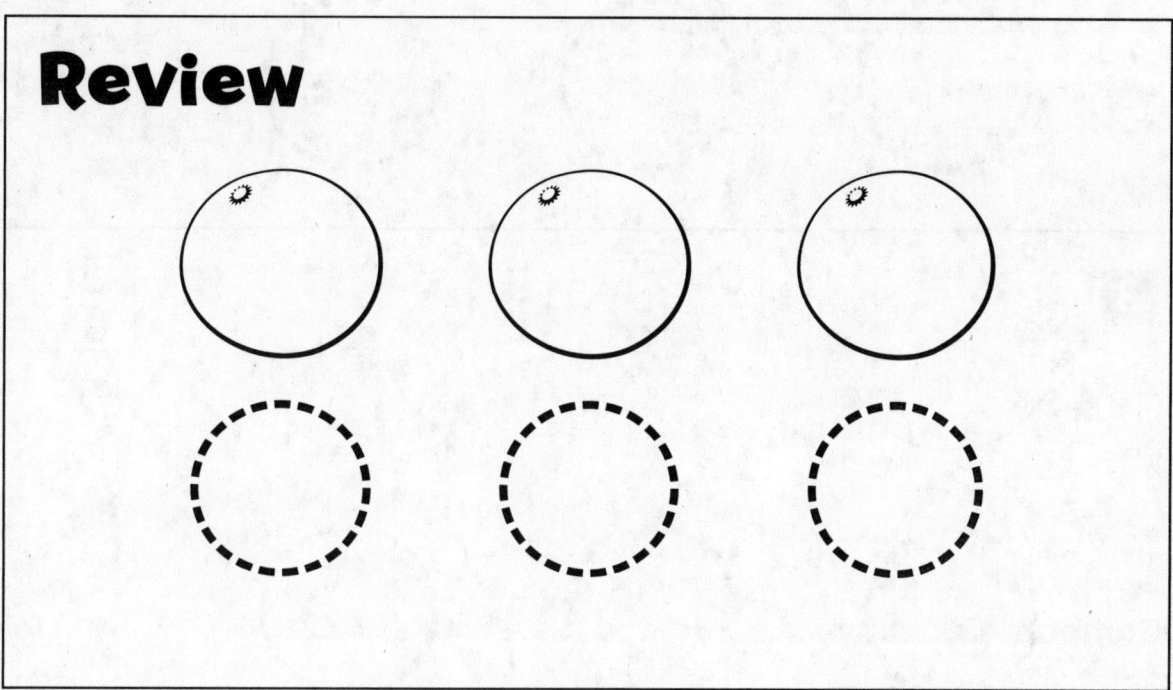

Review

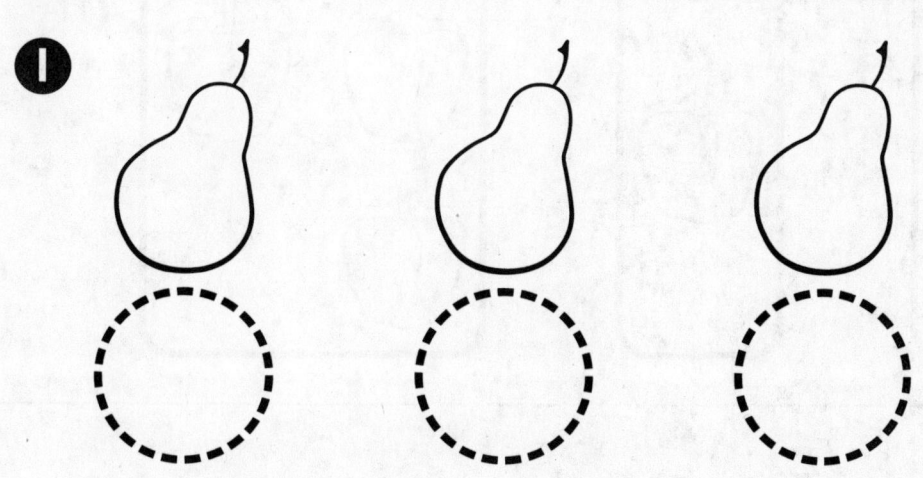

❶

Review: You can count the number of oranges.

Directions: 1. How many pears? Say how many. Color the counters to show how many.

Student Practice Book

1

2

○ ○ ○

3

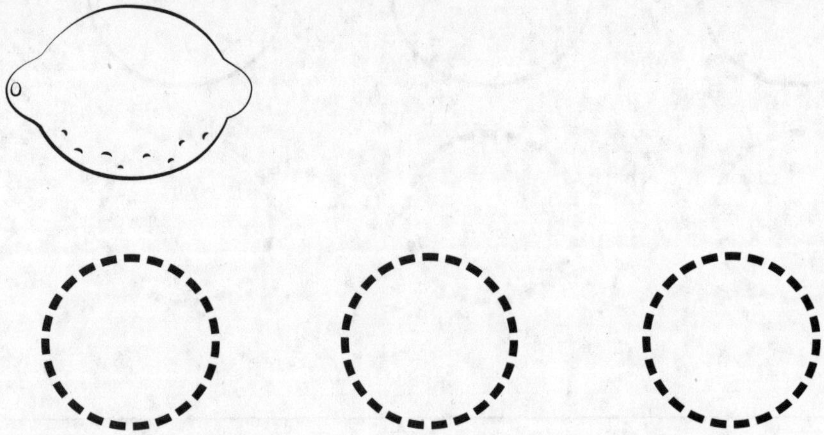

○ ○ ○

4

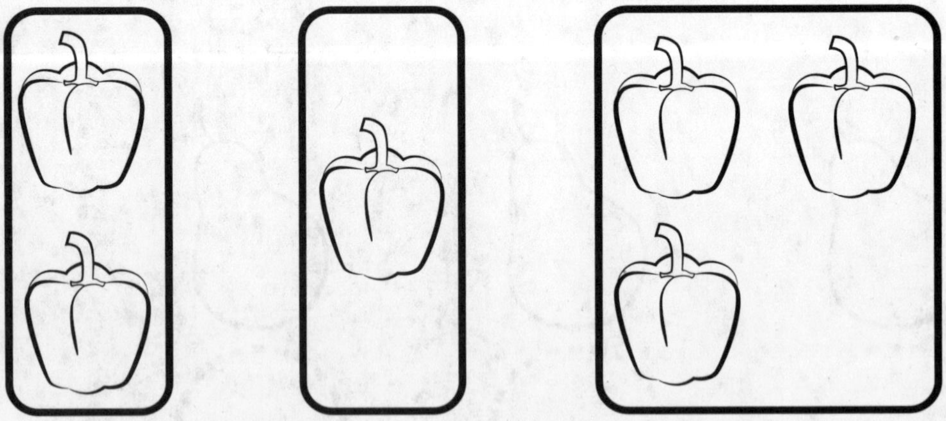

Copyright © McGraw-Hill Education

Math @ Home Activity

Give your child many opportunities to count everyday objects at home. Make a row of 2 or 3 objects, such as cans of food, cups, or apples. Have your child count the objects. Then ask how many there are.

Directions: 2. How many peaches? Color the counters to show how many. **3.** How many lemons? Color the counters to show how many. **4.** Circle the group with 3 peppers.

Student Practice Book

2

Additional Practice

Name _____

Review

1 2 ③ | ① 2 3

1

1 2 3

Review: You can count the number of apples and show how many are in each group.

Directions: 1. How many carrots? Circle the number to show how many.

Student Practice Book

3

2

1   2   3

3

1   2   3

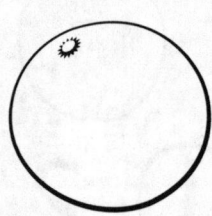

4

Directions: 2. How many peanuts? Circle the number to show how many. **3.** How many oranges? Circle the number to show how many. **4.** Draw 3 lemons.

Lesson 2-3

Additional Practice

Name _____

Review

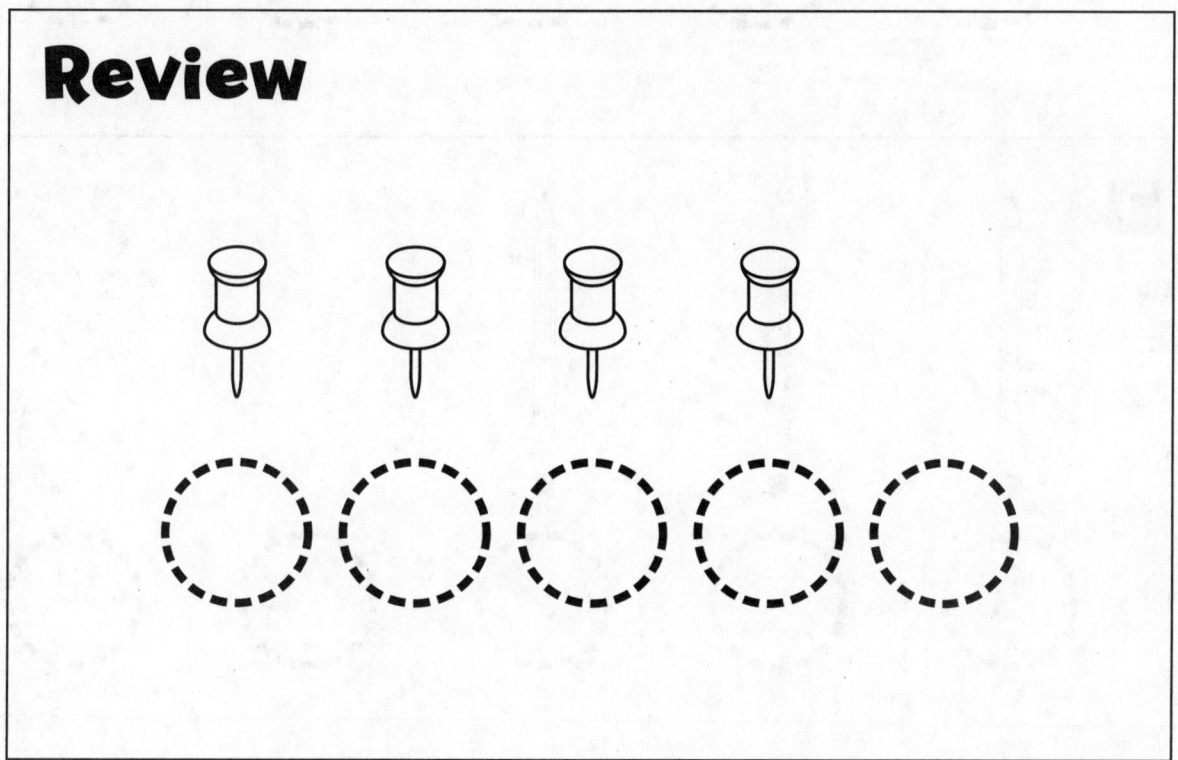

①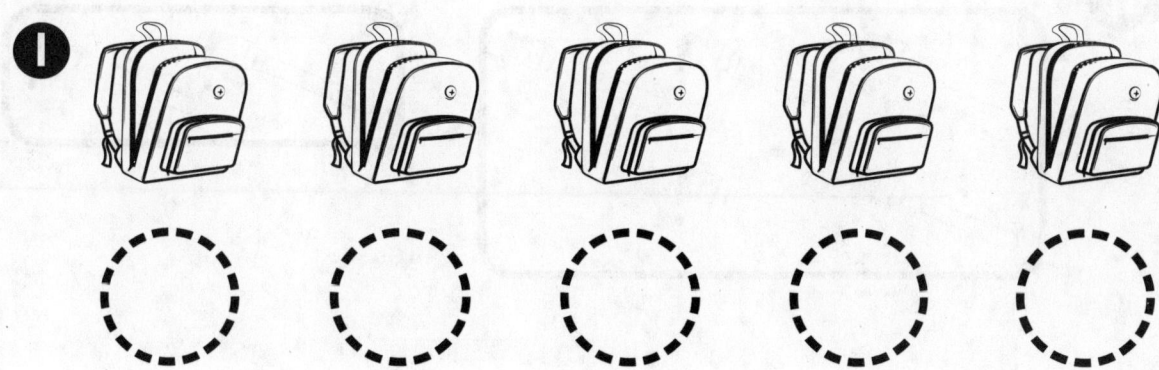

Review: You can count and say how many pushpins.

Directions: 1. How many backpacks? Say how many. Color the counters to show how many.

Student Practice Book

2 ⏺ 🔲🔲🔲🔲

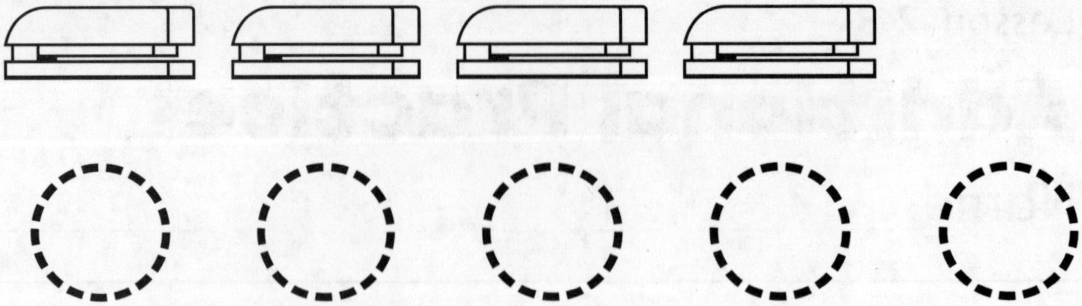

◯ ◯ ◯ ◯ ◯

---

3 ▪ ✏️✏️✏️

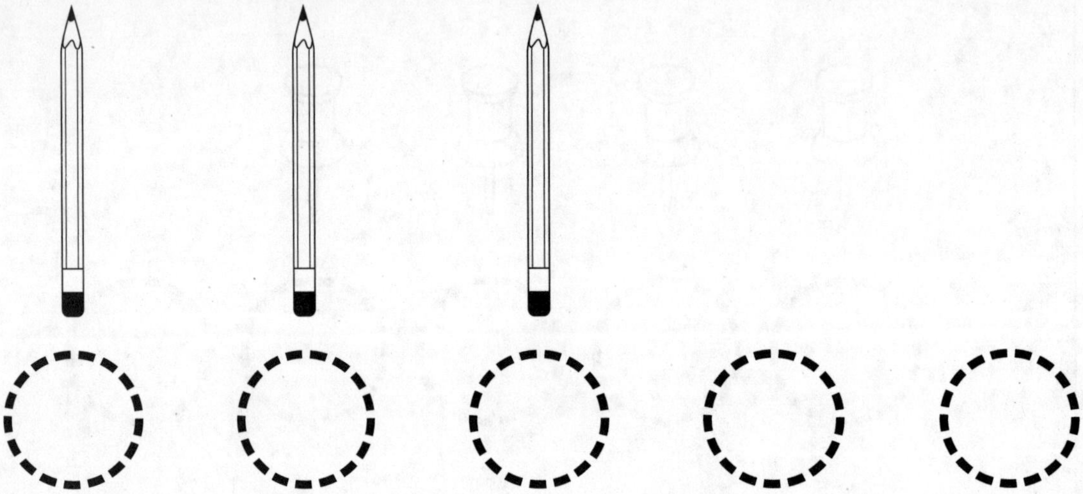

◯ ◯ ◯ ◯ ◯

---

4 🐟

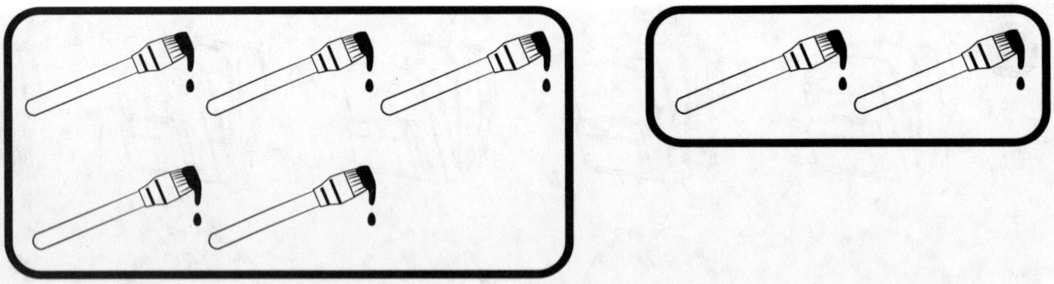

Give your child opportunities to count objects outside. Identify groups of 1–5 objects outside and ask your child to count and identify the number of objects in each group.

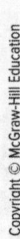

Directions: 2–3. How many objects? Color the counters to show how many. **4.** Which group has 5 paintbrushes? Circle the group with 5.

Student Practice Book

6

Lesson 2-4

Additional Practice

Name ________________________________

Review

5 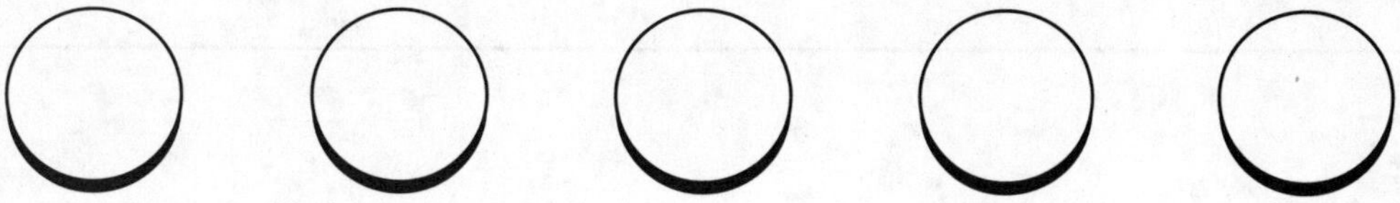

1

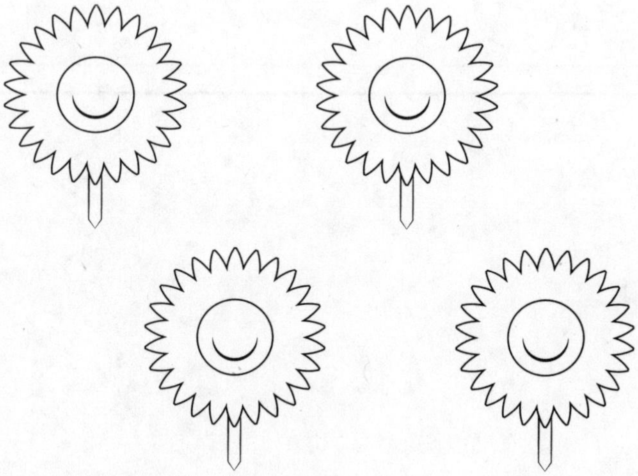

I 2 3 4 5

Review: You can use counters to show the number.

Directions: I. How many flowers? Circle the number to show how many.

2

1 2 3 4 5

3

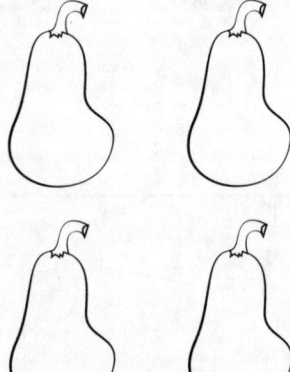

1 2 3 4 5

4

Directions: 2–3. How many objects? Circle the number to show how many. **4.** Draw 5 leaves.

Additional Practice

Name _____

Review

①

0 1 2 3 4 5

0 1 2 3 4 5

②

0 1 2 3 4 5

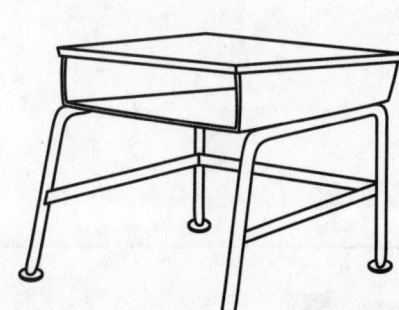

0 1 2 3 4 5

Review: You can tell how many birds are in the picture. There are 0 birds in the picture.

Directions: 1. How many pieces of chalk? Circle the number to show how many. **2.** How many books? Circle the number to show how many.

3

0 1 2 3 4 5   **0 1 2 3 4 5**

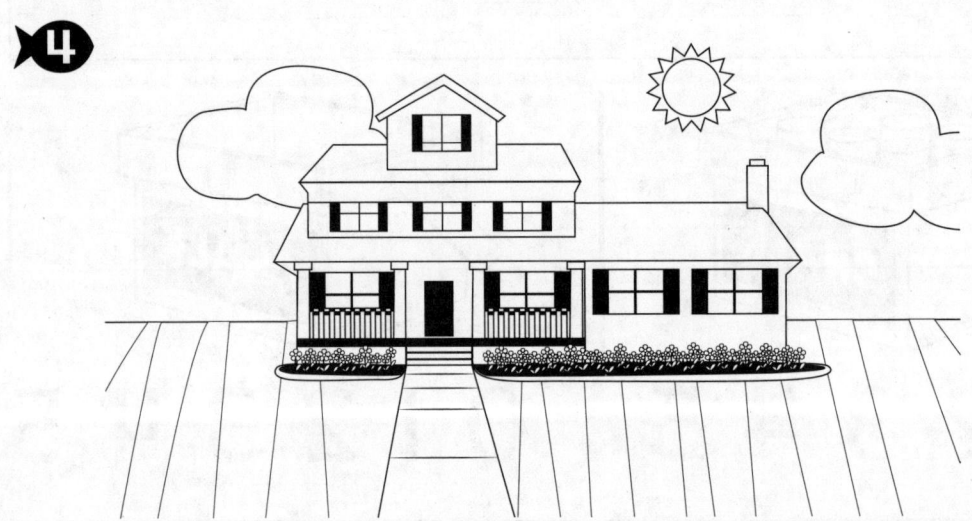 **4**

Math @ Home Activity

Look for situations around the house where you can ask your child to identify 0 objects. For example, if a grocery bag contains 5 onions, 4 carrots, and 3 potatoes, have your child identify how many bananas are in the bag.

Directions: 3. How many children? Circle the number to show how many. **4.** Draw 4 flowers and 0 trees by the house.

Additional Practice

Name _____

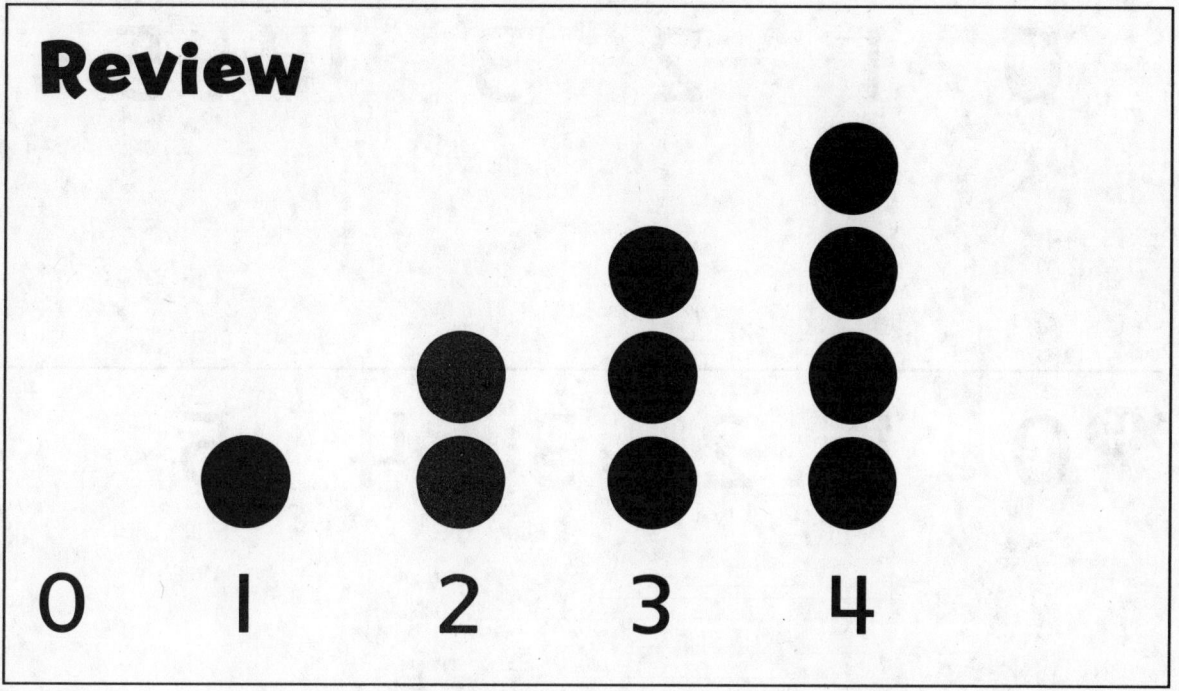

Review

0   1   2   3   4

①

②

Review: You can use counters to show how many.

Directions: 1. How many butterflies? Draw to show one more. **2.** How many spiders? Draw to show one more.

3

0   1   2   3   4   5

---

4 0   1   2   3   4   5

---

5 0   1   2   3   4   5

Math @ Home Activity

Give your child many opportunities to count everyday objects at home. Identify groups of 1-4 objects at home and ask your child to count the objects. Then ask how many there will be if you add one more.

Directions: 3. How many snails will there be if you add one more? Circle the number to show how many. **4.** What number is one more than 3? Circle the number that is one more. **5.** What number is one more than 4? Circle the number that is one more.

Student Practice Book

Lesson 2-7

Additional Practice

Name _____

Review

❶

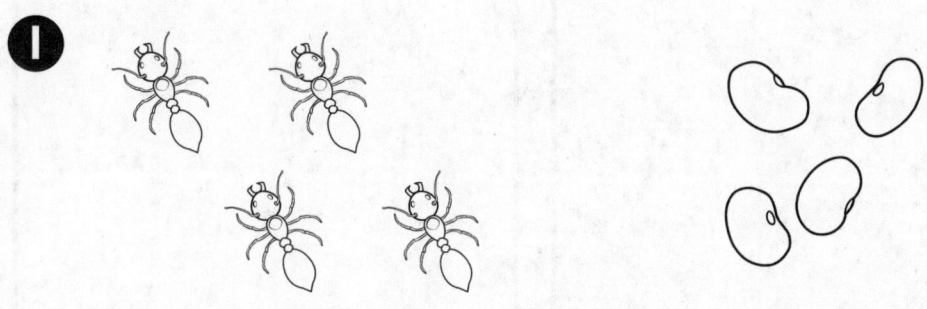

❷

Review: You can determine whether one group has the same number as the other group. There are the same number of bears as there are beavers.

Directions: 1–2. Are the two groups equal? Match the objects. Circle the groups if they are equal. Draw an X on the groups if they are not equal.

3

4

Math @ Home Activity

Give your child different opportunities to compare two groups of objects. Make different groups of toys and ask your child to compare two groups at a time. For example, have your child compare a group of 4 game pieces with a group of 2 game boards. Then have your child change one of the groups to make the two groups equal.

Directions: 3. Are the two groups equal? Match the objects. Circle the groups if they are equal. Draw an X on the groups if they are not equal. **4.** Draw to show groups that are equal.

Additional Practice

Name _____

Review

 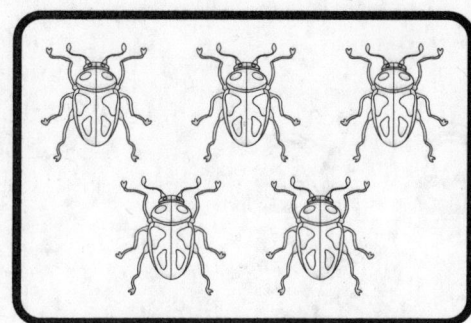

1

 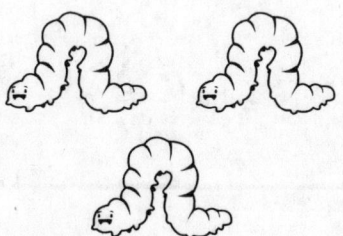

Review: You can compare groups using greater than or less than. The group of beetles is greater than the group of bees.

Directions: 1. Match objects in one group with objects in the other group. Circle the group that is greater than the other group. Draw an X on the group that is less than the other group.

2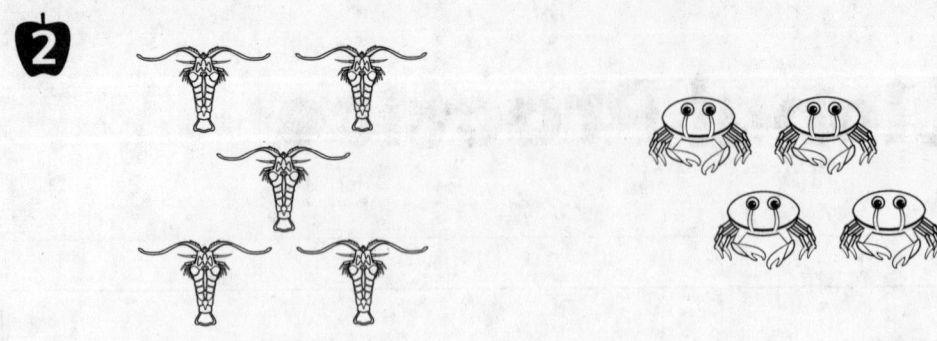

3

4

Math @ Home Activity

Provide opportunities for your child to compare groups using greater than or less than. For example, while folding laundry, ask your child to compare the different groups of clothing to one another.

Directions: 2–3. How can you compare the groups? Match objects in the groups. Circle the group that is greater than the other group. Draw an X on the group that is less than the other group. **4.** Count the fish. Draw a group of fish that is greater.

Student Practice Book

Lesson 2-9

Additional Practice

Name _____

Review

3 ▲▲▲

5 ▲▲▲▲▲

3 is *less than* 5.

1

5 [　　　　　　]

4 [　　　　　　]

Review: You can draw objects to show each number. Then, you can compare the groups of objects to find which is greater than, less than, or equal to.

Directions: 1. Which number is greater? Draw objects to show each number. Circle the number that is greater.

Student Practice Book

17

Copyright © McGraw-Hill Education

2

3 [   ]

1 [   ]

3

5 [   ]

5 [   ]

4

4   2

1   5   3

Math @ Home Activity

Provide your child with opportunities to compare groups of up to 5 everyday objects at home. Make several rows of the same type of object, with each row containing a different number of that object. Have your child compare the number of objects in the different rows.

Directions: 2. Which number is less? Draw objects to show each number. Circle the number that is less. **3.** Draw objects to show equal groups. **4.** Which numbers are less than 3? Circle all of the numbers that are less than 3.

Additional Practice

Name _____

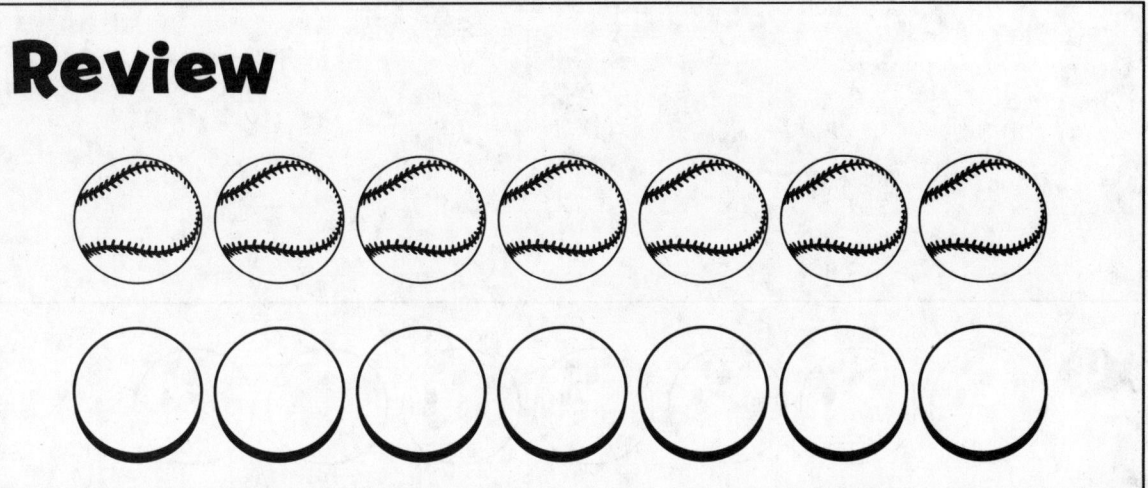

Review

1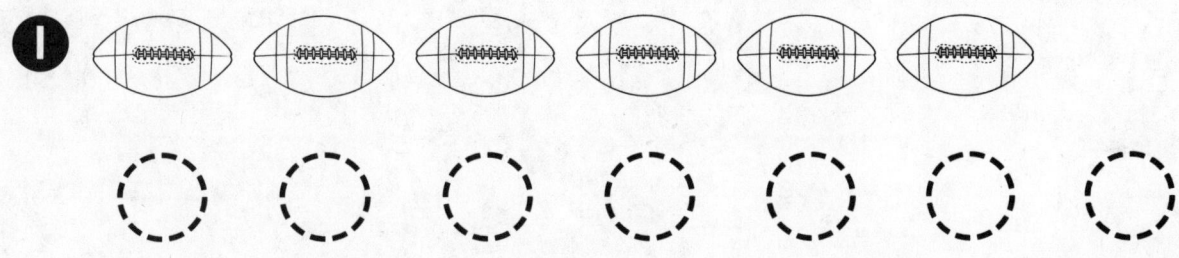

2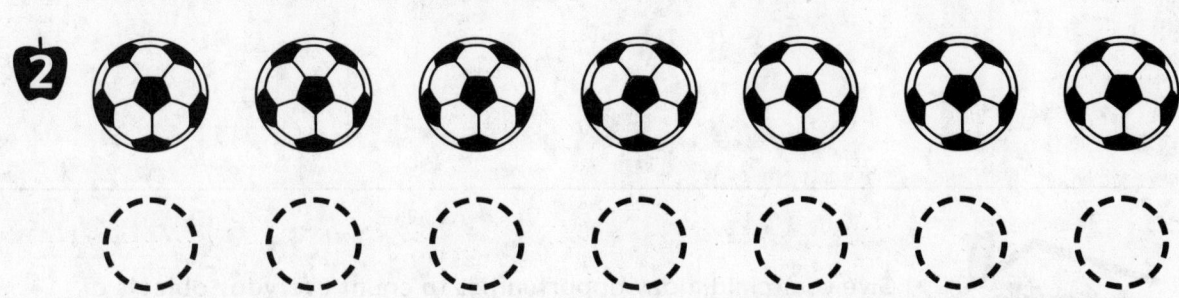

Review: You can count objects. The last number you say tells how many.

Directions: 1–2. How many balls? Color the counters to show how many.

3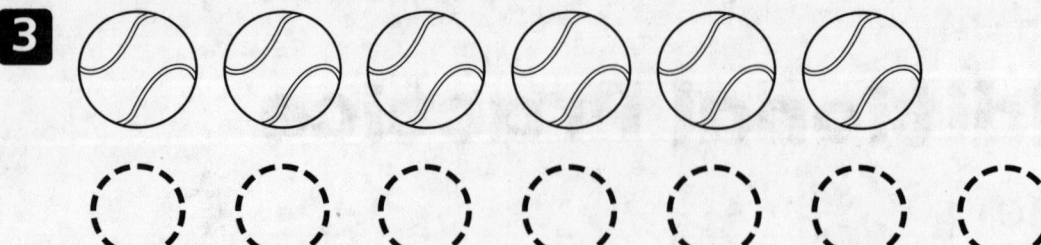

4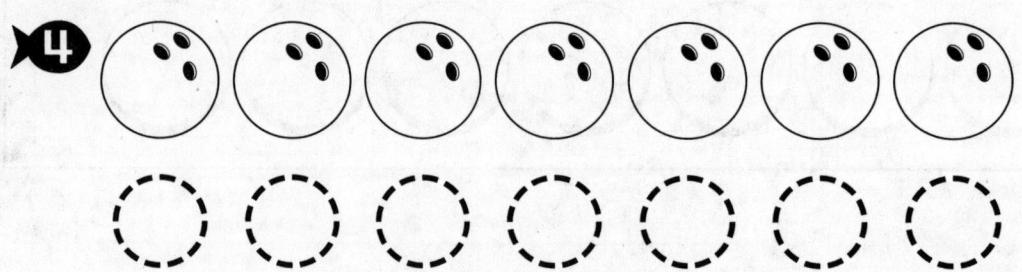

Math @ Home Activity

Give your child many opportunities to count everyday objects at home. While cleaning up toys, have your child count aloud as each toy is put away. After counting to 7, have your child start counting over again, beginning at 1.

Directions: 3–4. How many balls? Color the counters to show how many.

Lesson 3-2

Additional Practice

Name _____

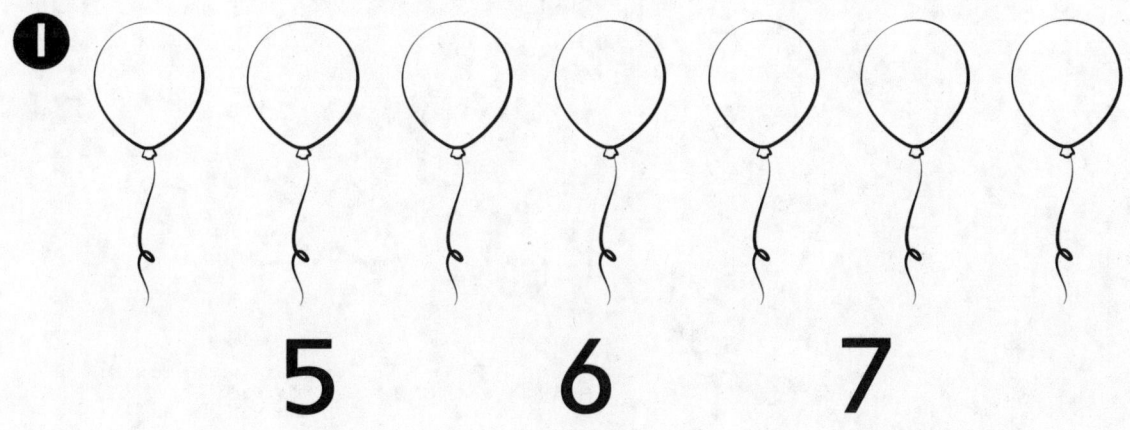

Review: You can show, or represent, numbers in different ways. You can show 6 or 7.

Directions: 1–2. How many objects? Circle the number to show how many.

3

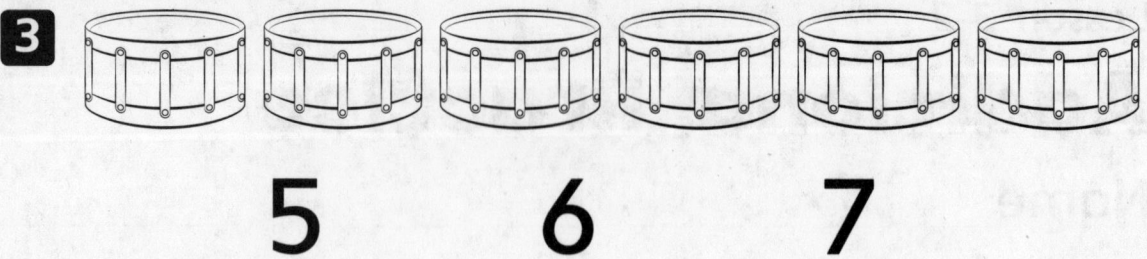

5   6   7

4

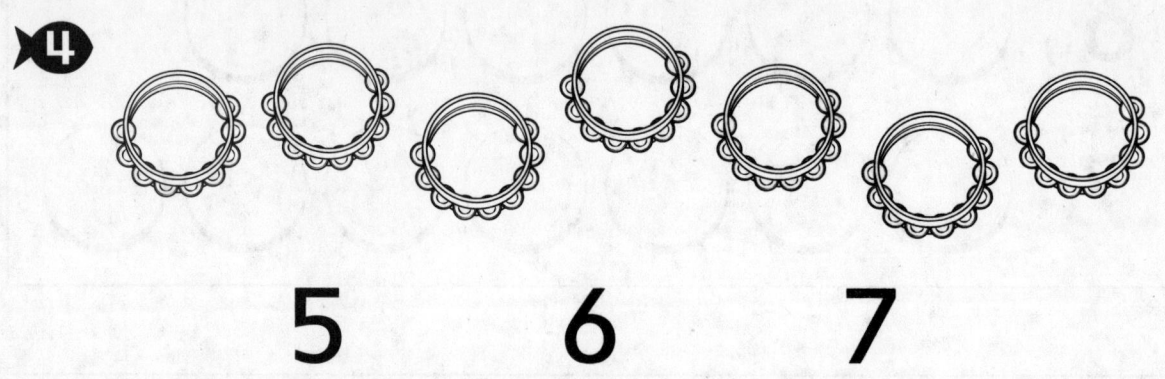

5   6   7

Math @ Home Activity

Practice showing 6 and 7 with your child. Using stickers, toothpicks, or paper clips, make groups of 6 and 7. Have your child count the number of objects, and then say how many out loud.

Directions: 3–4. How many objects? Circle the number to show how many.

Lesson 3-3
Additional Practice

Name _____

Review

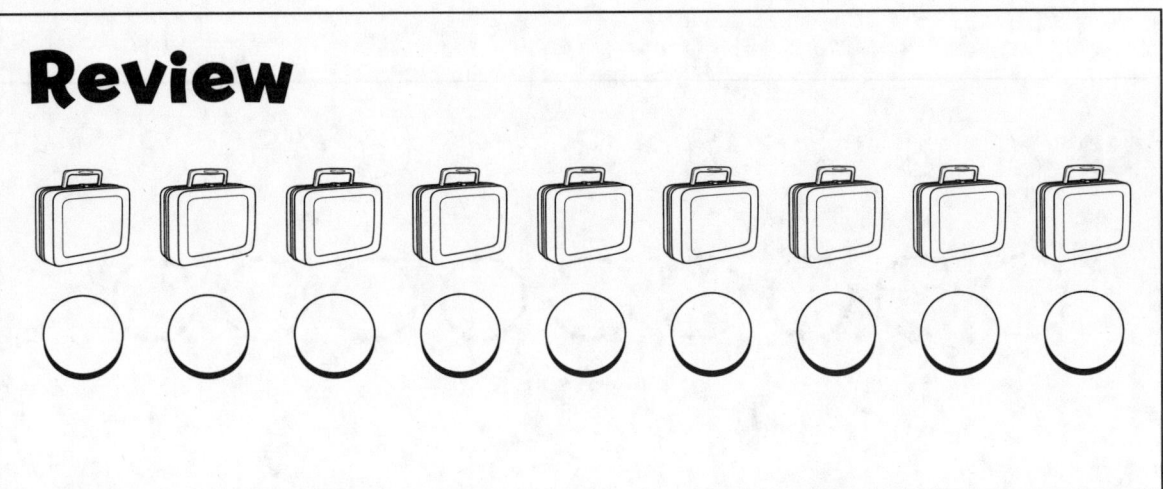

1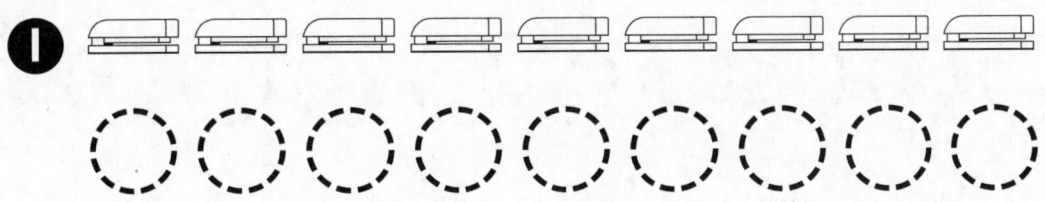

2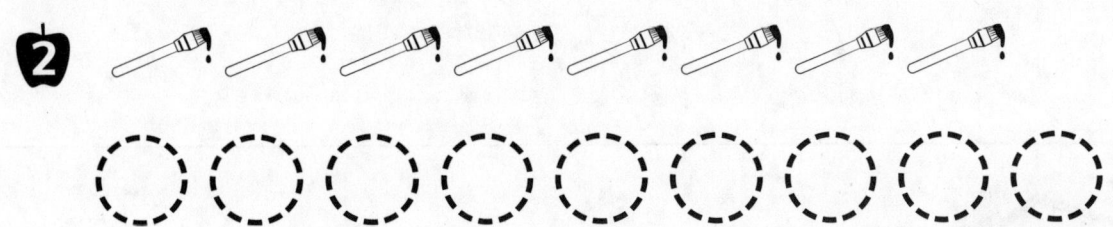

Review: You can count objects. When you count, the last number you say tells how many.

Directions: 1–2. How many objects? Color the counters to show how many.

3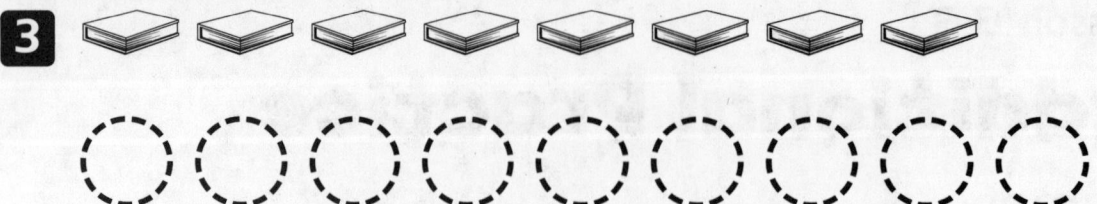

○ ○ ○ ○ ○ ○ ○ ○ ○

4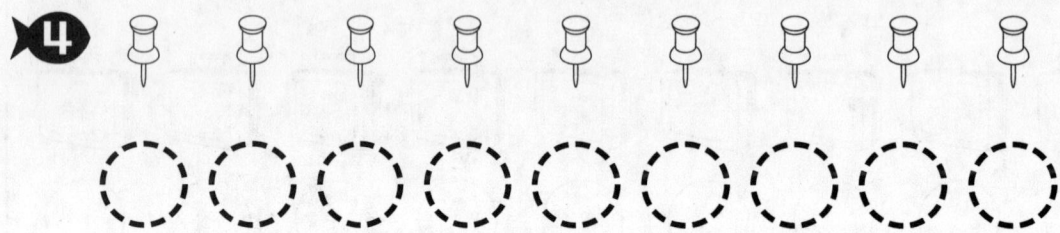

○ ○ ○ ○ ○ ○ ○ ○ ○

5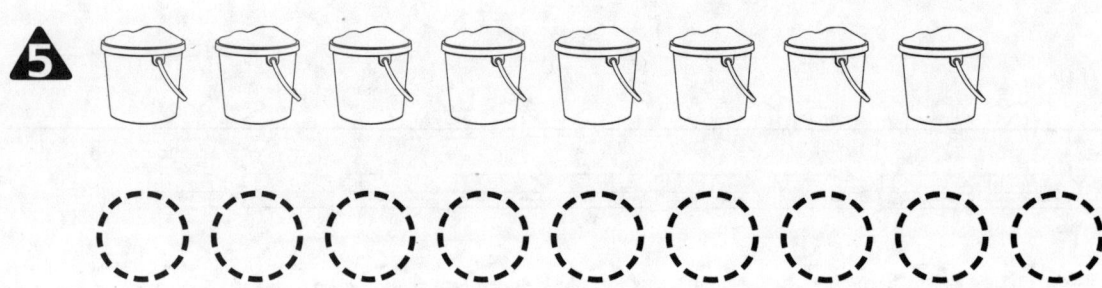

○ ○ ○ ○ ○ ○ ○ ○ ○

Math @ Home Activity

Using finger paints or a stamp pad, have your child make groups of fingerprints. Count with your child to determine how many fingerprints are in each group.

Directions: 3–5. How many objects? Color the counters to show how many.

Additional Practice

Name _____

Review

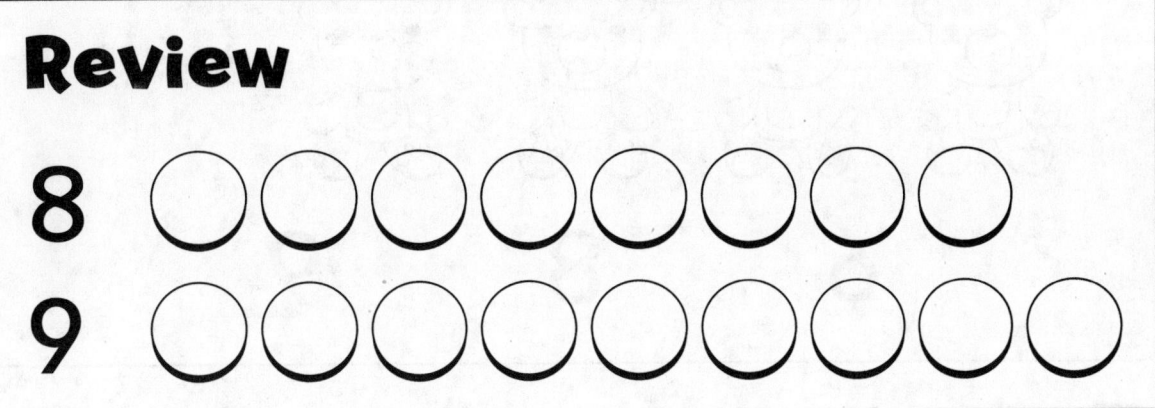

8

9

1

7   8   9

Review: You can use a number to show, or represent, how many. You can show 8 or 9.

Directions: 1. How many toys? Circle the number to show how many.

2

6   8   9

3

5   6   8

4

5   8   9

Directions: 2–4. How many objects? Circle the number to show how many.

Lesson 3-5

Additional Practice

Name _____

Review

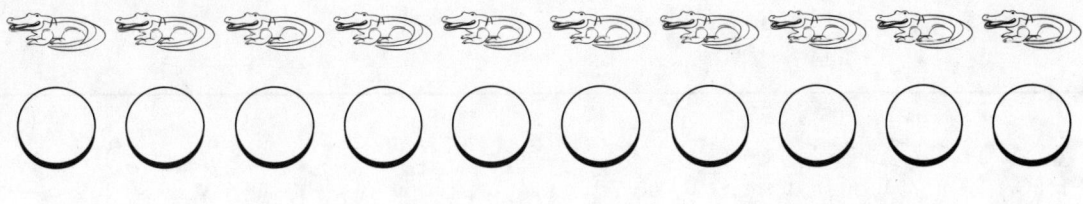

1

2

Review: You can use counters to help you count objects. The last number you say tells how many.

Directions: 1–2. How many animals? Color the counters to show how many.

3

○ ○ ○ ○ ○ ○ ○ ○

4

○ ○ ○ ○ ○ ○ ○ ○ ○ ○

Math @ Home Activity

Look for opportunities around your home where your child can practice counting to 10. Using pennies or other small objects, make groups of 1 to 10. Have your child count aloud.

Directions: 3–4. How many animals? Color the counters to show how many.

Lesson 3-6

Additional Practice

Name ....

Review

10 ○ ○ ○ ○ ○ ○ ○ ○ ○ ○

1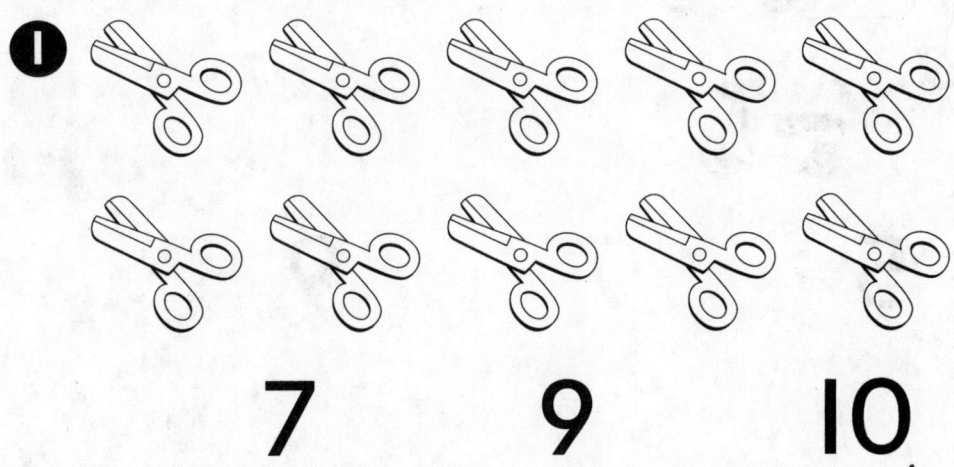

7   9   10

2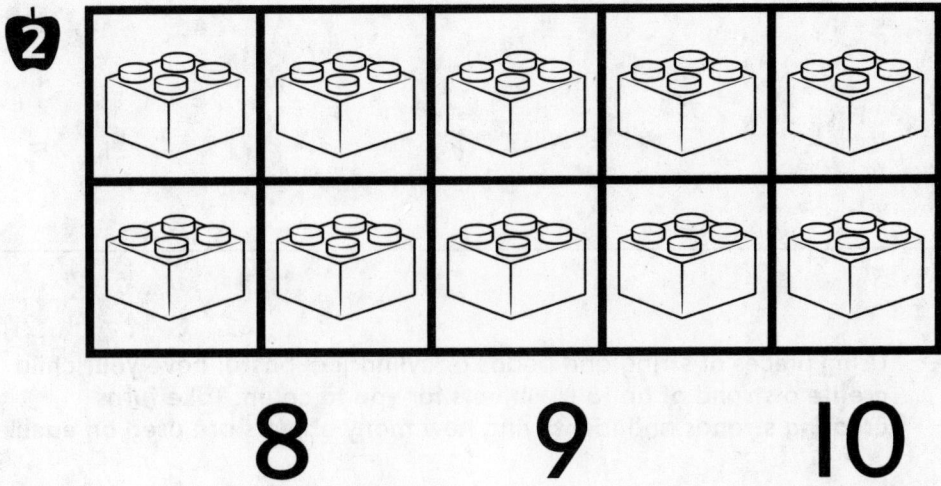

8   9   10

Review: You can show, or represent, numbers in different ways. You can show 10.

Directions: 1–2. How many objects? Circle the number to show how many.

3

8   9   10

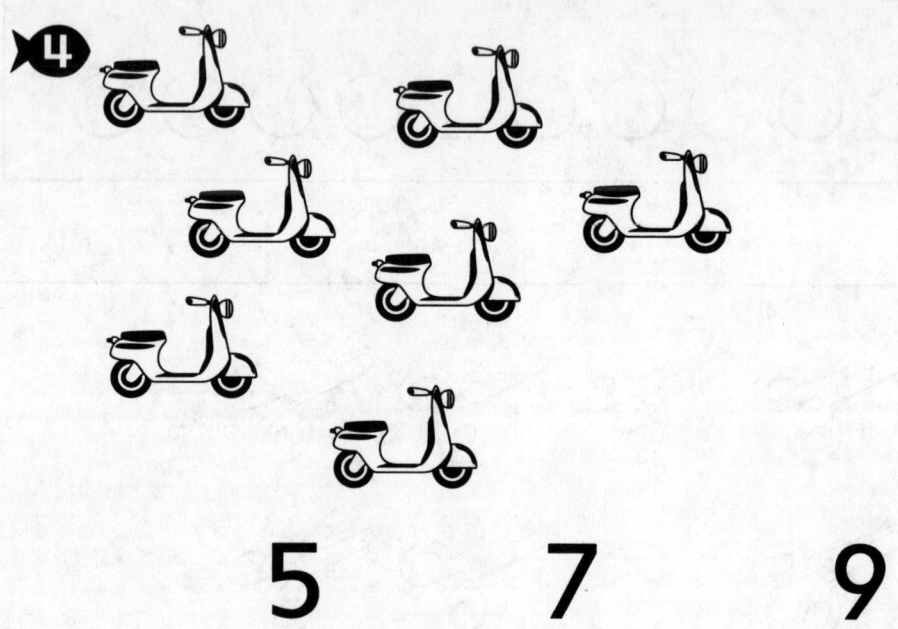

5   7   9

Math @Home Activity

Using pieces of string and beads or cylindrical pasta, have your child create a strand of up to 10 objects for you to count. Take turns creating strands and identifying how many objects are used on each.

Directions: 3–4. How many objects? Circle the number to show how many.

Additional Practice

Name

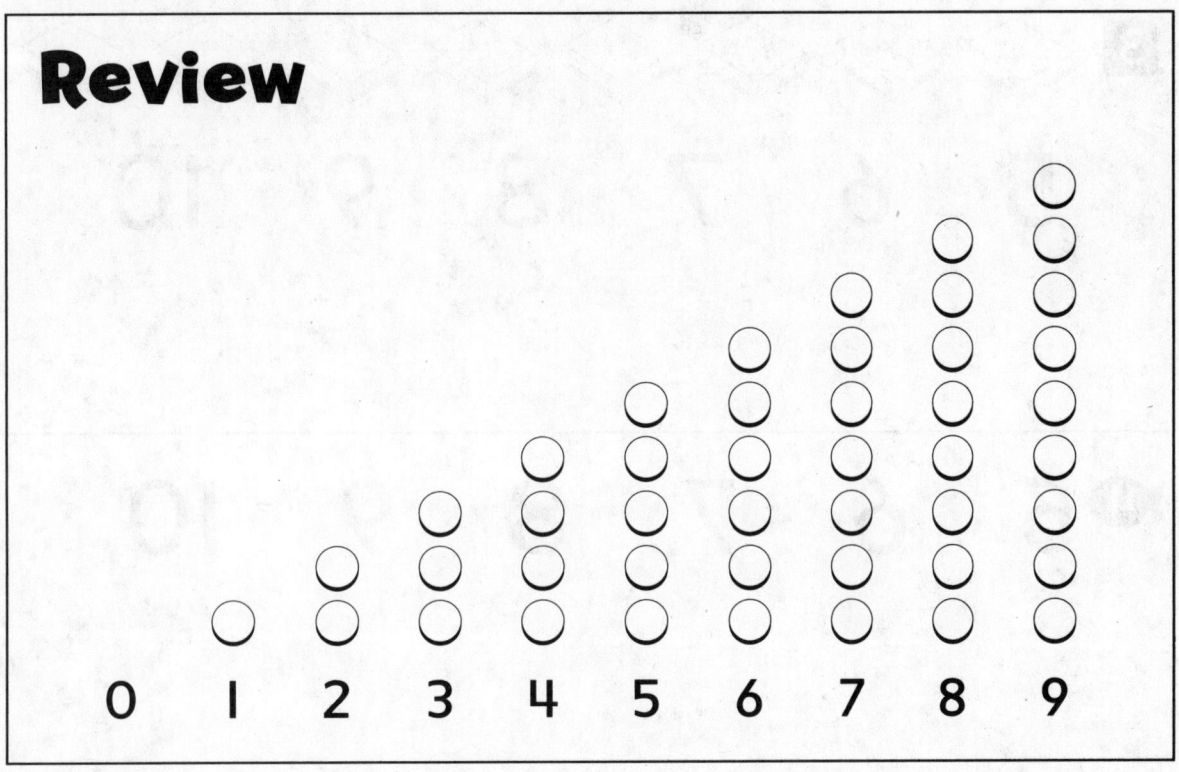

Review

0 1 2 3 4 5 6 7 8 9

1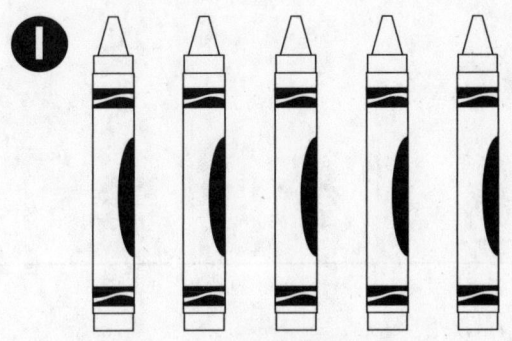

Review: You can use counters to show one more.

Directions: 1. How many crayons? Draw to show one more.

2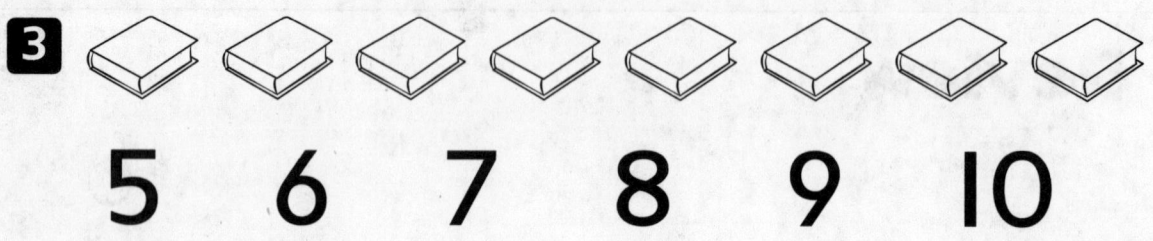

3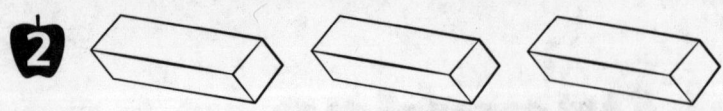

5 6 7 8 9 10

---

4 5 6 7 8 9 10

---

5 5 6 7 8 9 10

Math @ Home Activity

Give your child many opportunities to count everyday objects at home. Identify groups of 1–9 objects at home and ask your child to count the objects. Then ask how many there will be if you add one more.

Directions: 2. How many erasers? Draw to show one more. **3.** How many books will there be if you add one more? Circle a number to show how many. **4.** Circle the number that is one more than 7. **5.** Circle the number that is one more than 9.

Additional Practice

Name

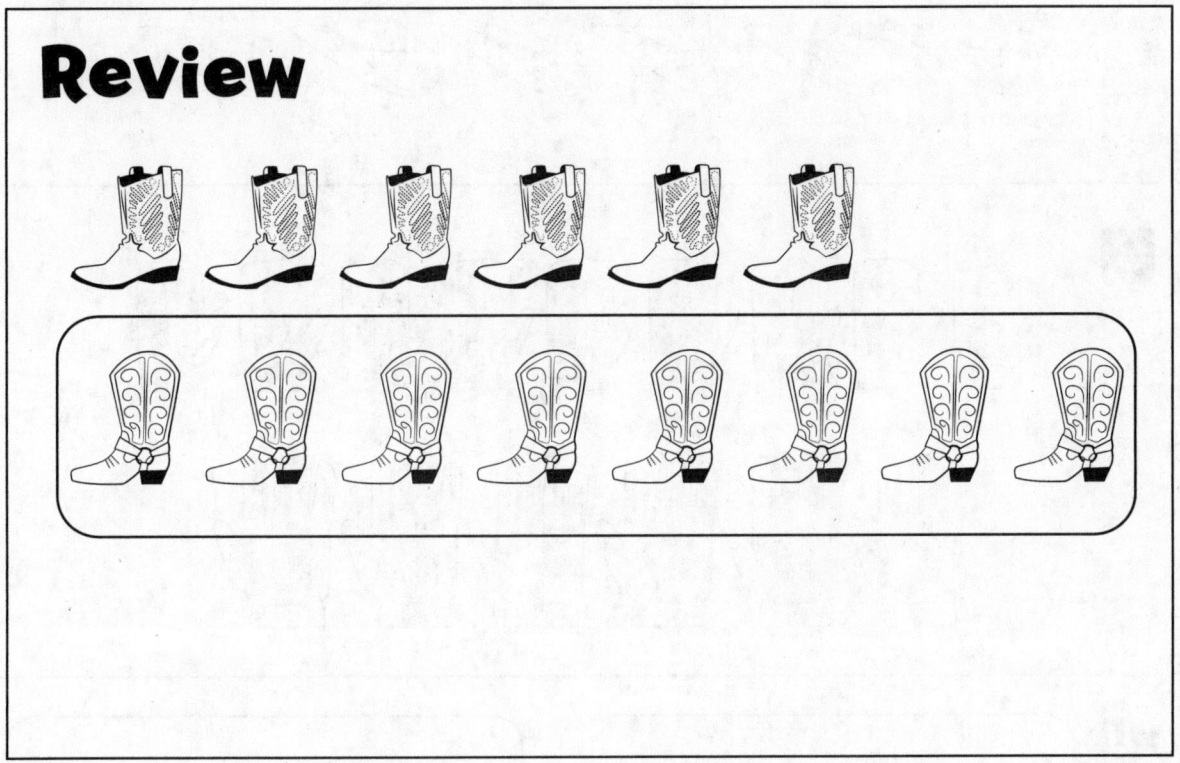

1

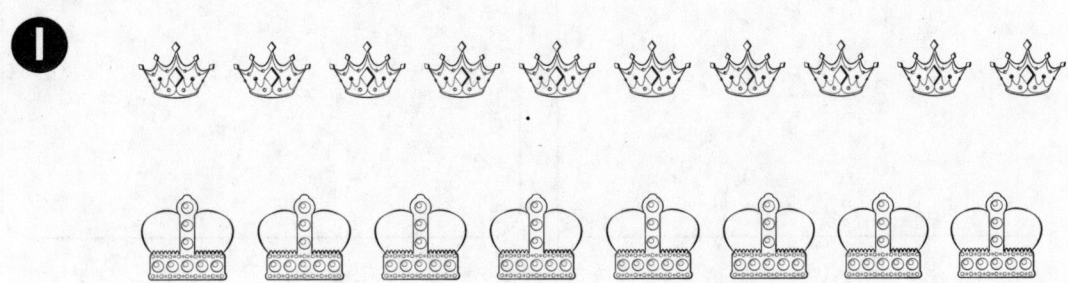

Review: You can compare two groups to find out which has more or fewer, or if the groups are equal. The group with 8 boots has more than the group with 6 boots.

Directions: 1. How can you compare the two groups? Circle the group that is greater. Put an X on the group that is less. Circle both groups if they are equal.

2

3

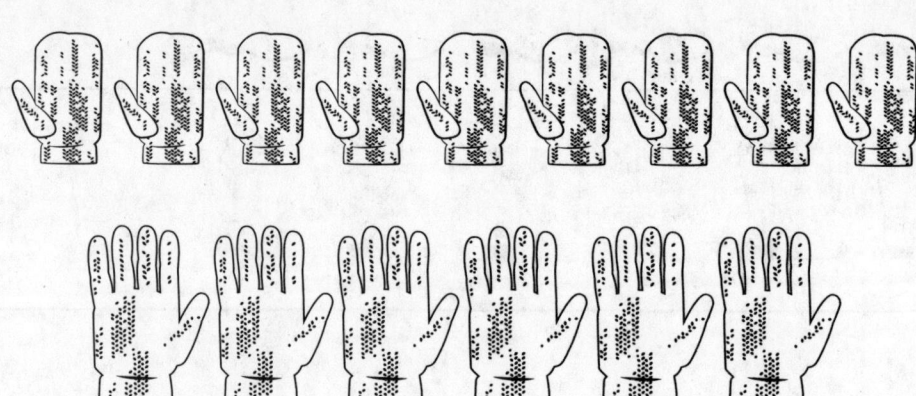

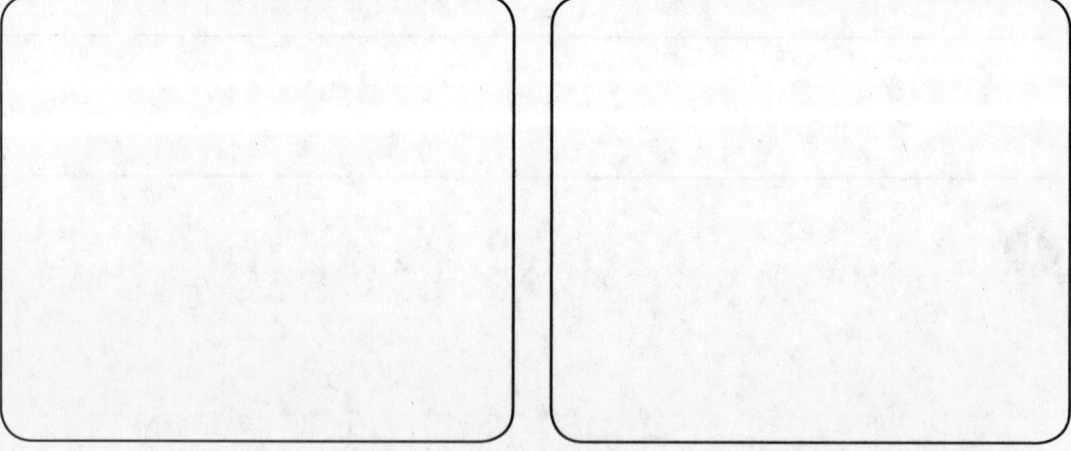

Math @ Home Activity

Give your child many opportunities to compare groups with more, fewer, or an equal number of objects. While playing outside, have your child make groups of rocks or leaves. Have your child determine which group has more or fewer objects, or whether the groups contain the same number of objects. Ask your child to explain their reasoning.

Directions: 2–3. How can you compare the two groups? Circle the group that is greater. Put an X on the group that is less. Circle both groups if they are equal. **4.** How can you draw to show equal groups?.

Additional Practice

Name _____

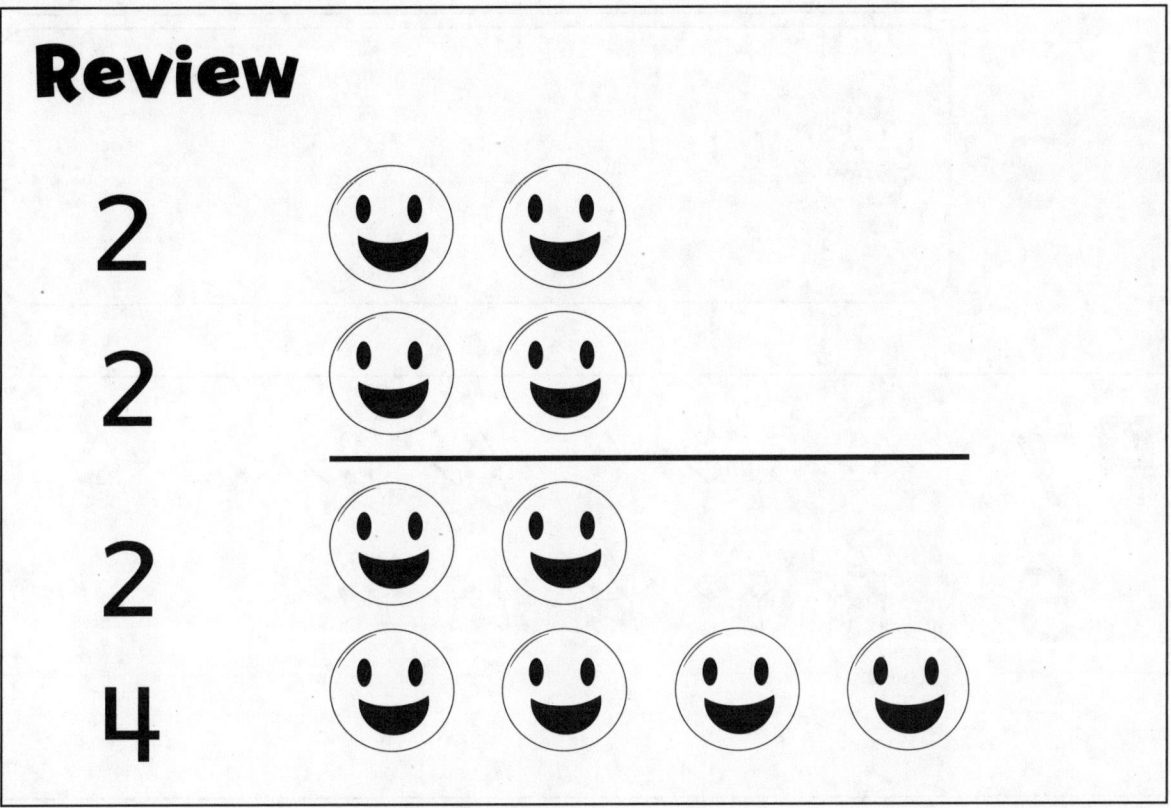

Review

2
2
___
2
4

❶ 4

6

Review: You can compare numbers by counting how many objects are in each group.

Directions: I. How can you draw to show each number? Circle the number that is greater.

2

5

3

3 7

5

4 1 2 3 4 5

Math @ Home Activity

Give your child many opportunities to compare groups to find greater or less. Have your child make groups of pennies or paperclips. Count each group. Have him or her determine which group is greater or less. Ask your child to explain his or her reasoning.

Directions: 2. How can you draw to show each number? Circle the number that is less. **3.** Which number is greater? Circle it. **4.** Circle all of the numbers that are less than 3.

Additional Practice

Name _____

Review

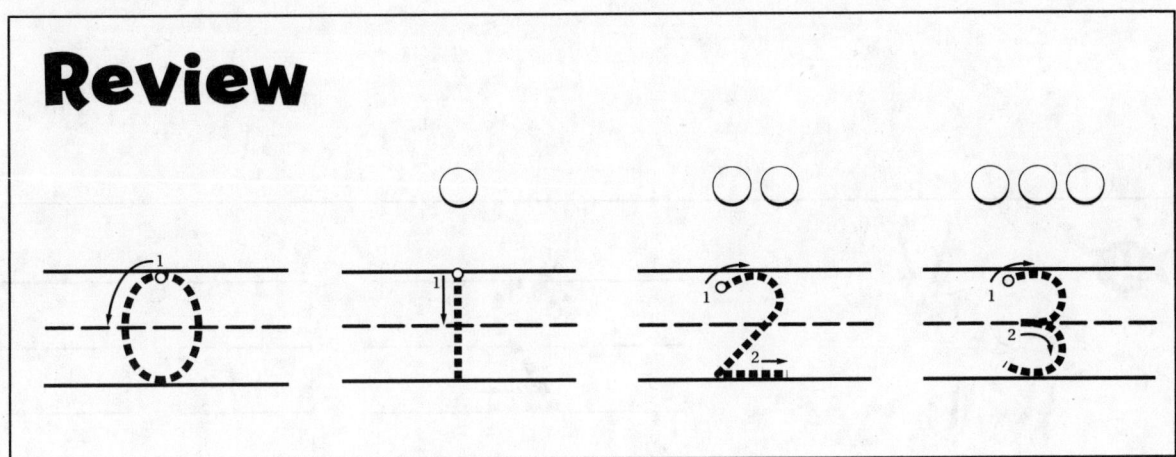

1

2
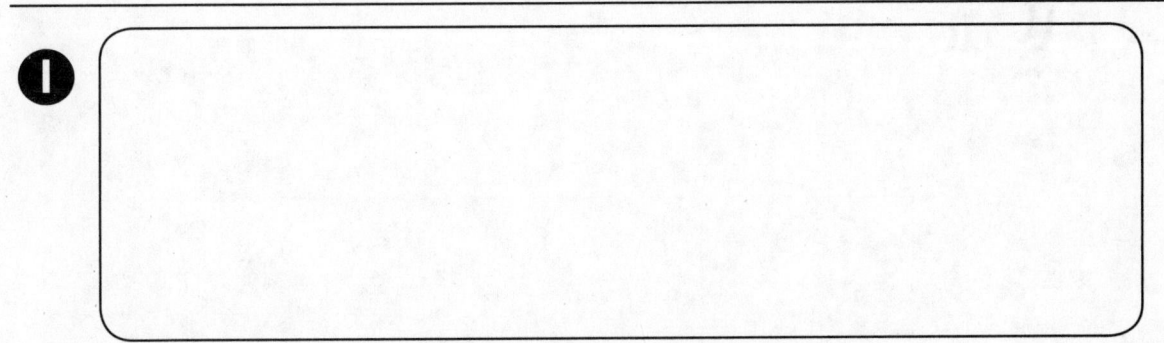

Review: You can trace and write the numbers 0, 1, 2, and 3 to show how many.

Directions: 1. Write two 3s in the box. **2.** Trace and write the number with a pencil. Then, trace the pencil with different colored crayons to practice.

Student Practice Book

3

 4

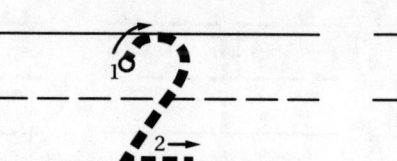

Copyright © McGraw-Hill Education

 **Math @ Home Activity**

On a cookie sheet or plate filled with sand, rice, or shaving cream, have your child practice writing the numbers 0 to 3. Show your child a number of objects and have him or her write that number in the sand.

Directions: 3–4. How many animals? Trace and write the number with a pencil. Then, trace the pencil with different colored crayons to practice.

Additional Practice

Name _____

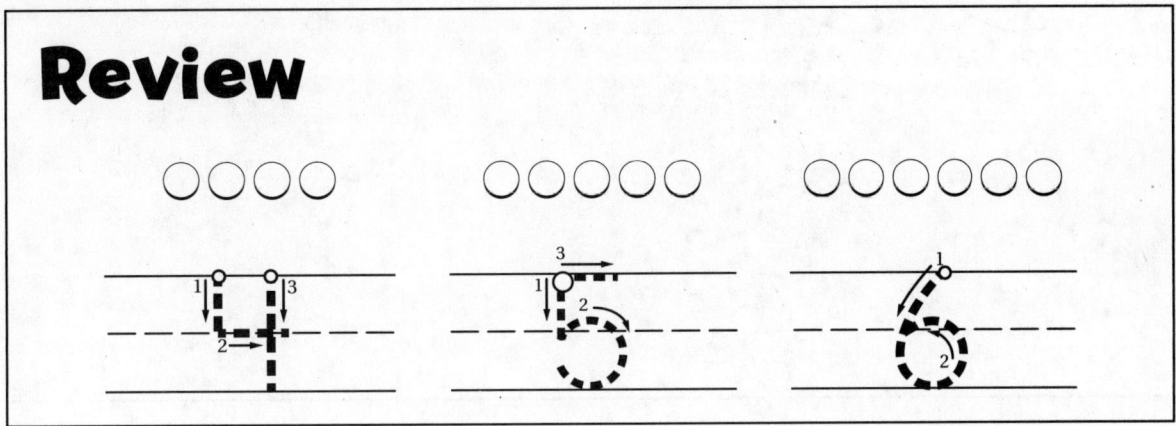

Review

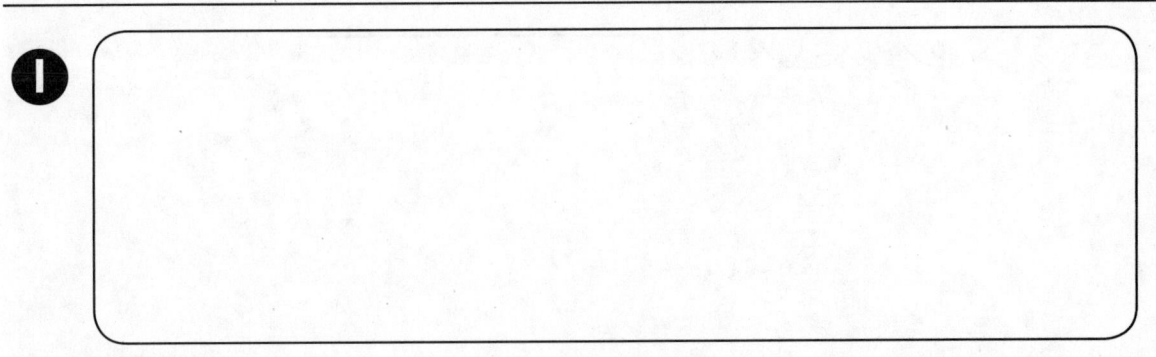

1

2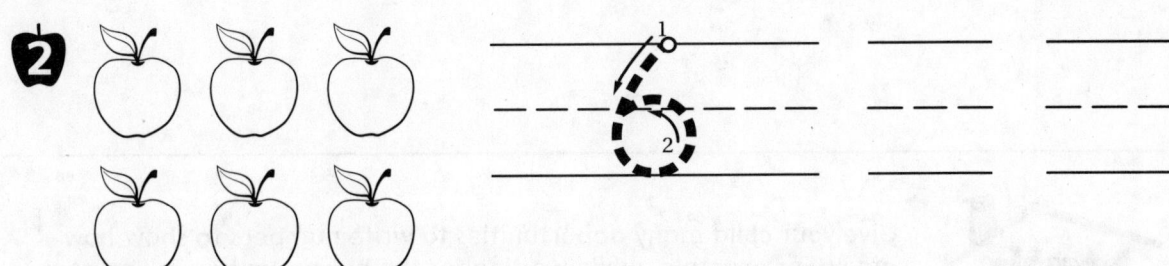

Review: You can trace and write the numbers 4, 5, and 6 to show how many.

Directions: 1. Write five 5s in the box. **2.** How many apples? Trace and write the number with a pencil. Then, trace the pencil with different color crayons to practice.

3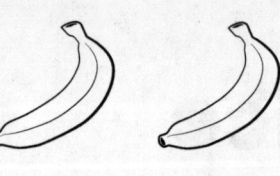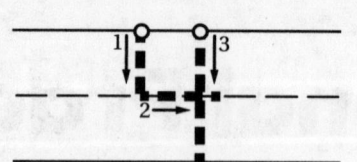

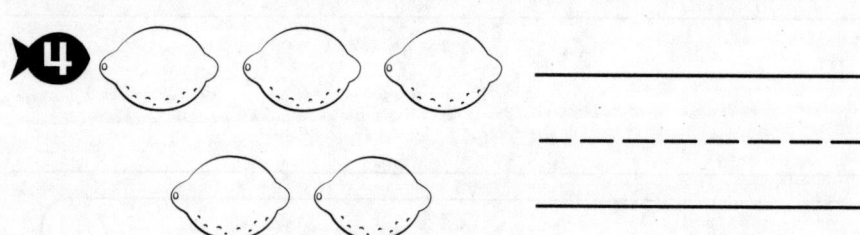

4

Math @ Home Activity

Give your child many opportunities to write numbers to show how many. For example, while waiting for the bus or waiting in line at the grocery store, say a number from 4 to 6 out loud. Have your child repeat the number, count that number of nearby objects, and then write the number in the air with their finger.

Directions: 3. How many bananas? Trace and write the number with a pencil. Then, trace the pencil with different color crayons to practice. **4.** How many lemons? Write the number.

Additional Practice

Name _____

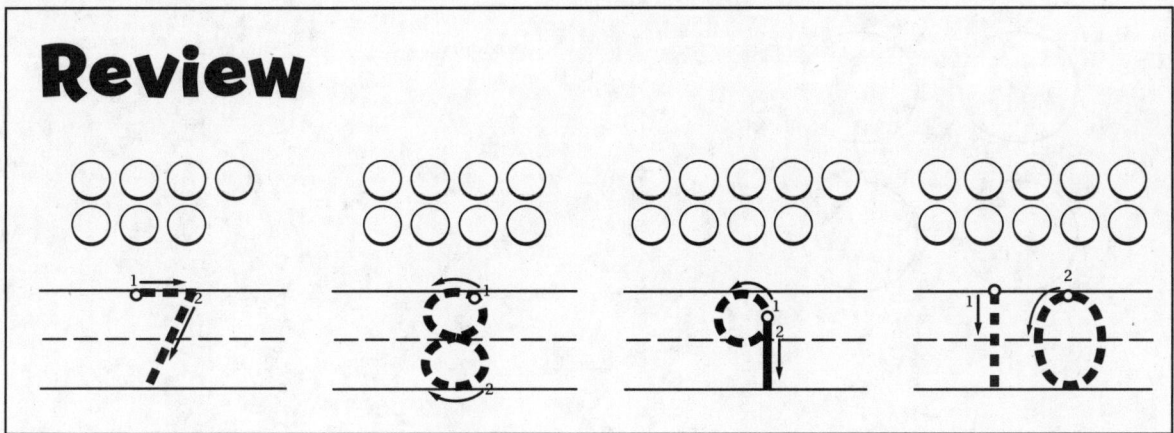

Review

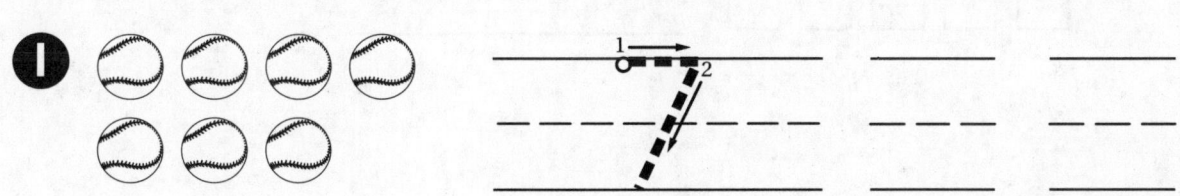

1

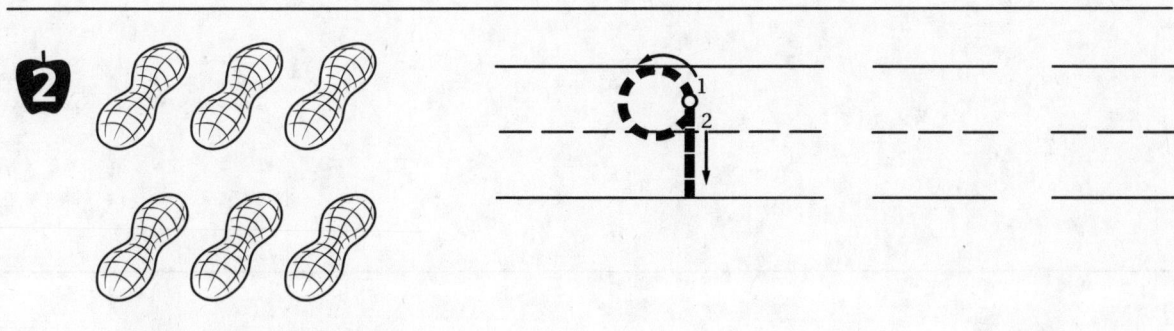

2

Review: You can trace and write the numbers 7, 8, 9, and 10 to show how many.

Directions: 1–2. How many objects? Trace and write the number with a pencil. Then, trace the pencil with different color crayons to practice.

3

4

_____

- - - - - - - - - - -

_____

Directions: 3. How many oranges? Trace and write the number with a pencil. Then, trace the pencil with different color crayons to practice. **4.** Color the squares. How many squares? Write the number.

Additional Practice

Name _____

Review

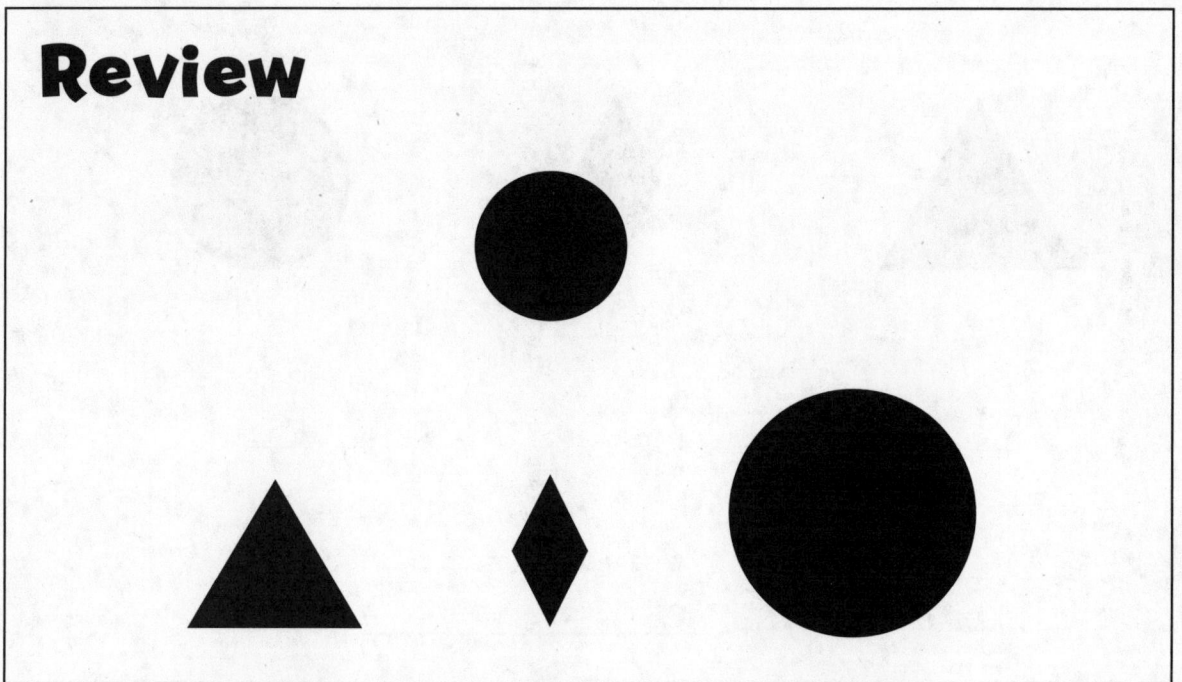

❶

❷

Review: You can find differences between objects. The shape that is different from the shape above in only one way has an X on it.

Directions: 1–2. Which object is different? Draw an X on the object that is different.

3

Directions: 3. Which shape has only one difference from the shape above? Circle it. Explain your thinking.

Additional Practice

Name _____

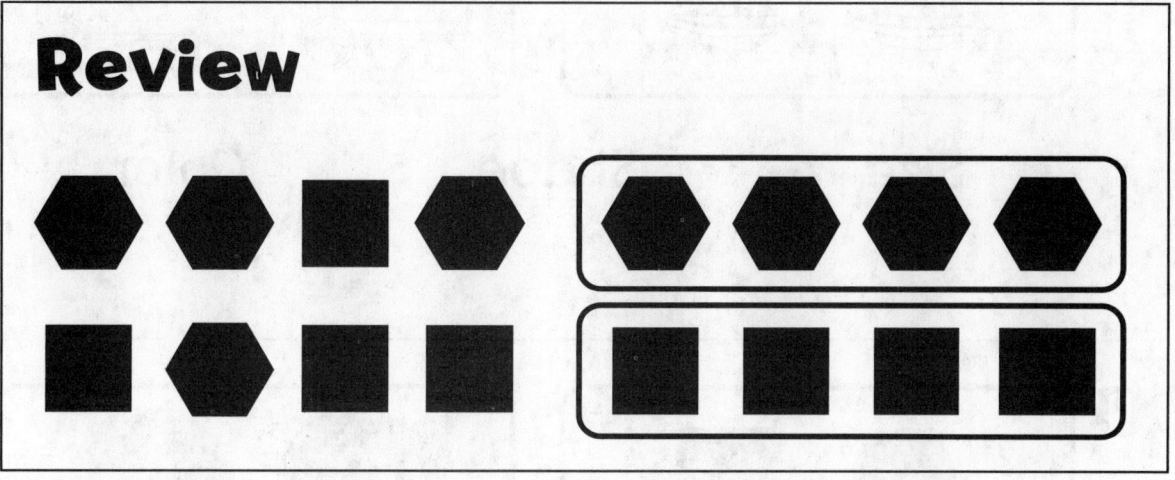

1.

2.

Review: You can sort objects into groups by color, size, or shape.

Directions: 1. How can you sort by shape? Draw a line from each shape to a box. **2.** How can you sort these objects? Draw a line from each object to a box.

Size   Shape   Color

4

Size   Shape   Color

Math @Home Activity

Provide many opportunities for your child to sort objects into groups based on identified attributes. For example, while sitting outside, ask your child to identify the characteristic that makes the leaves look different. If possible, gather leaves with your child and sort them into groups based on their color, shape, or size.

Directions: 3–4. How are the objects sorted? Circle to show if they are sorted by size, shape, or color.

Lesson 4-3

Additional Practice

Name _____

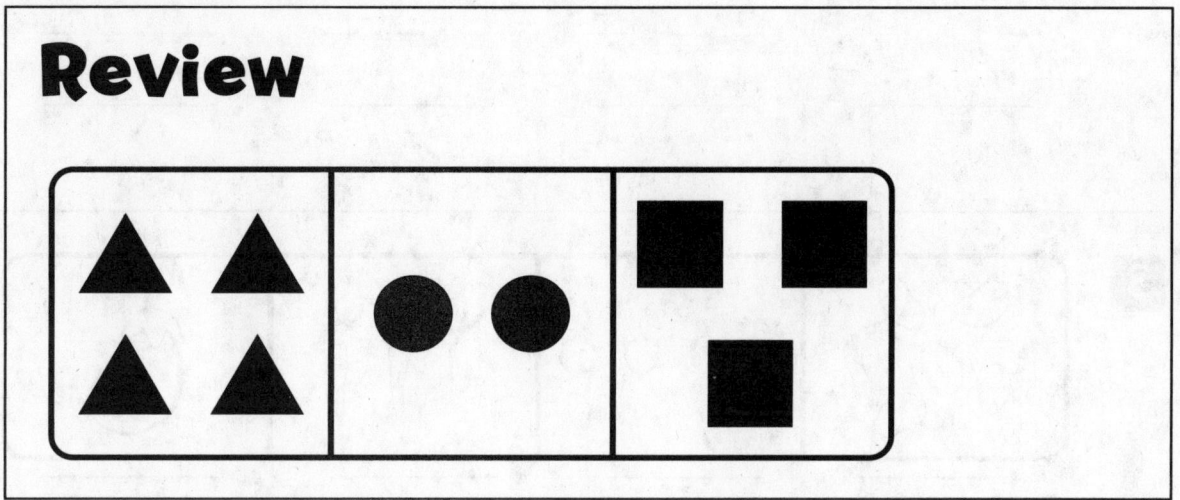

Review

1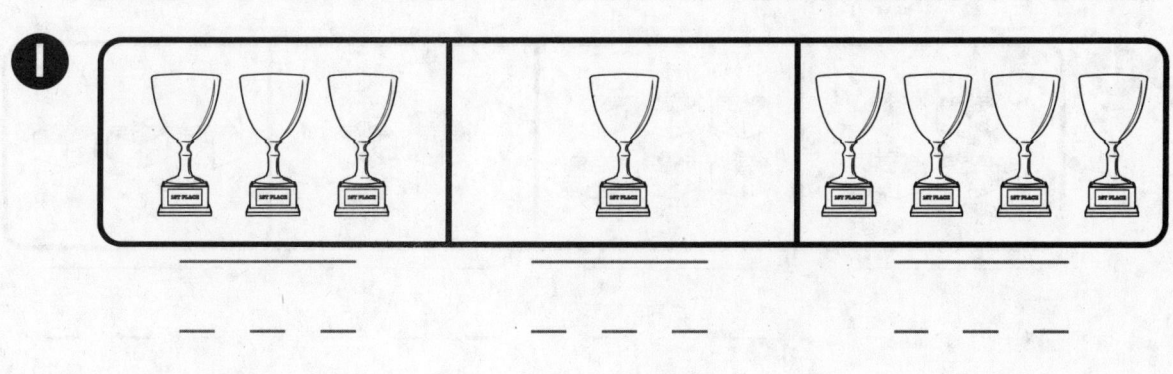

_ _ _ _   _ _ _ _   _ _ _ _

_____   _____   _____

Review: You can count how many objects are sorted into groups. There are 4 triangles, 2 circles, and 3 squares.

Direction: 1. How many objects are in each group? Write a number to show how many.

2

_____ _____ _____ _____

- - - - - - - - - - - - - - - -

_____ _____ _____ _____

3

_____ _____ _____ _____

- - - - - - - - - - - - - - - -

_____ _____ _____ _____

4

_____ _____ _____ _____

- - - - - - - - - - - - - - - -

_____ _____ _____ _____

Math @ Home Activity

Provide your child with many opportunities to identify the number of objects in different groups. For example, have your child sort marbles based on their color. Have your child count how many of each color marble are in each group.

Directions: 2–4. How many objects are in each group? Write a number to show how many.

Additional Practice

Name

Review

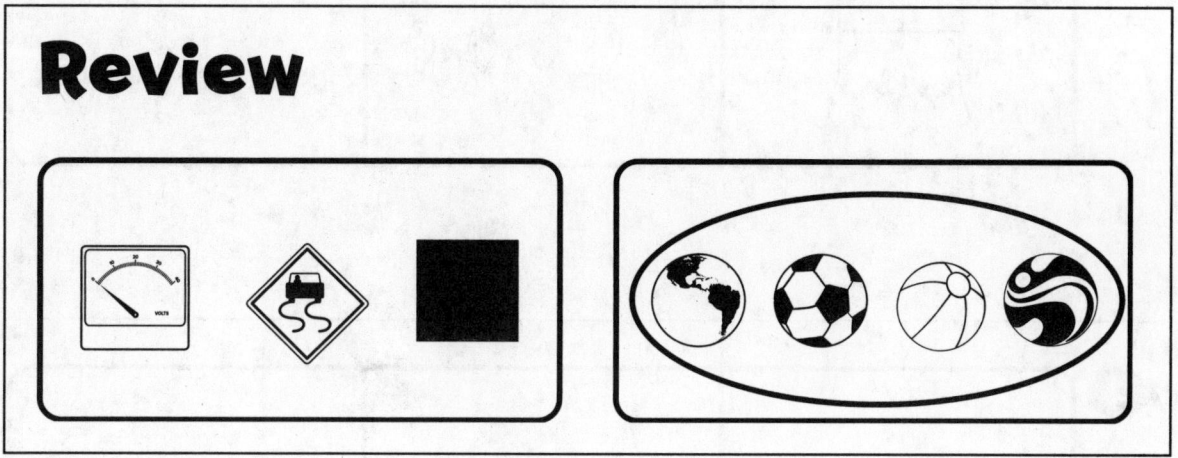

1

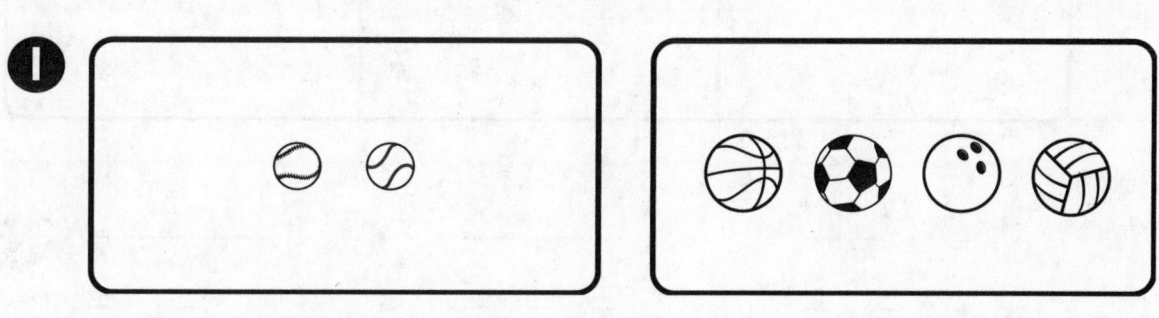

2

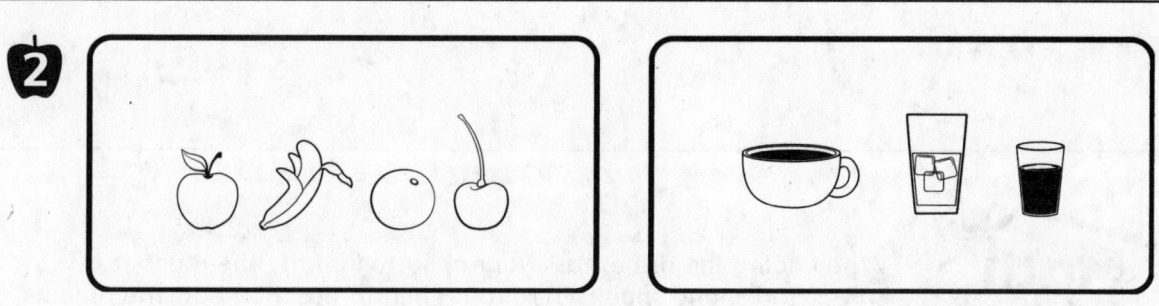

Review: You can sort objects into groups by their attributes. These objects are sorted by shape. 3 objects are square and 4 are round. The group of round objects has more.

Direction: 1. How many objects in each group? Circle the group that has more. **2.** How many objects in each group? Circle the group that has fewer.

3

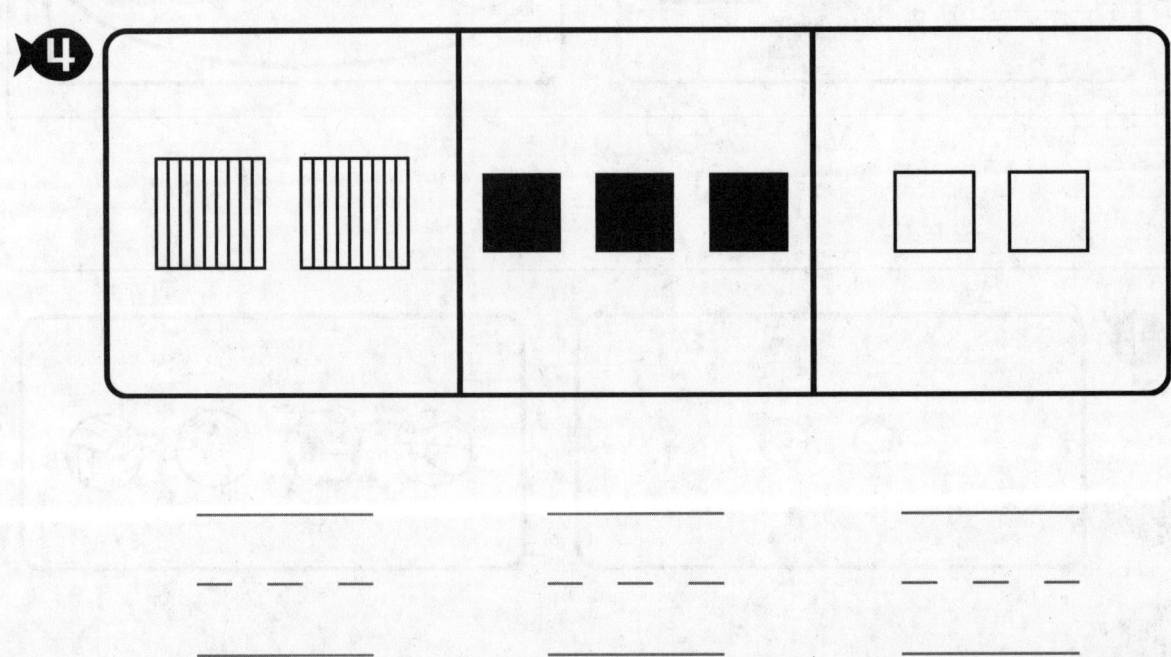

4

_ _ _   _ _ _   _ _ _

_____   _____   _____

Directions: 3. How many objects in each group? Circle the group that has the fewest. **4.** How many square objects in each group? Write numbers to show how many. Circle the groups that have the same number.

Additional Practice

Name _____

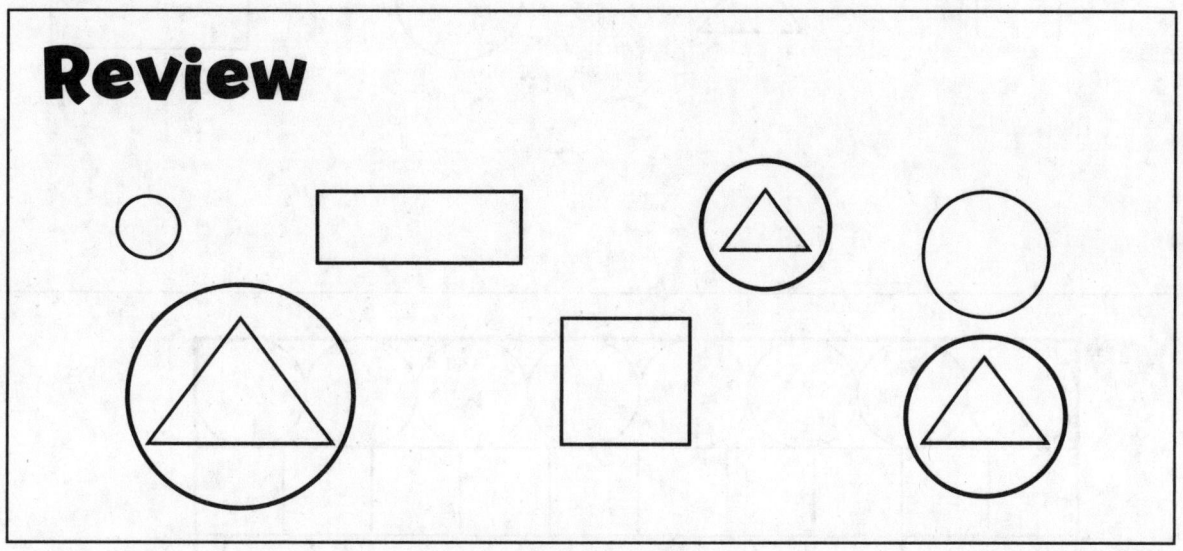

Review

---

❶

---

❷

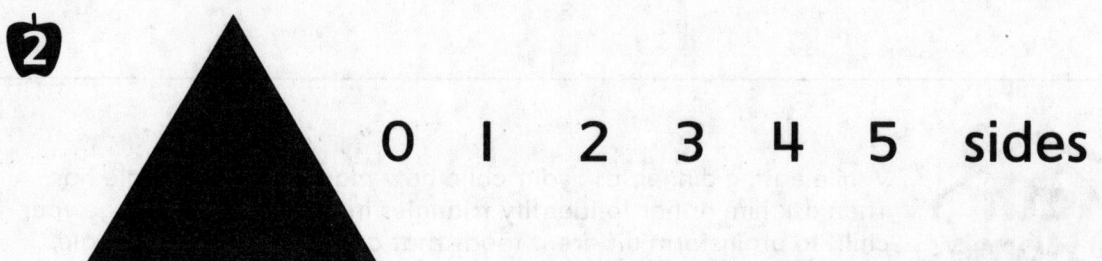

0   1   2   3   4   5   sides

Review: You can identify and name a triangle. A triangle has 3 sides.

Directions: 1. Which object is shaped like a triangle? Circle the object that is shaped like a triangle. **2.** How many sides? Circle the number to show how many.

Student Practice Book

51

3

 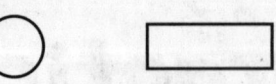  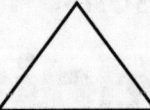

   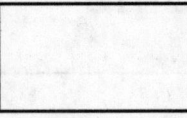

4

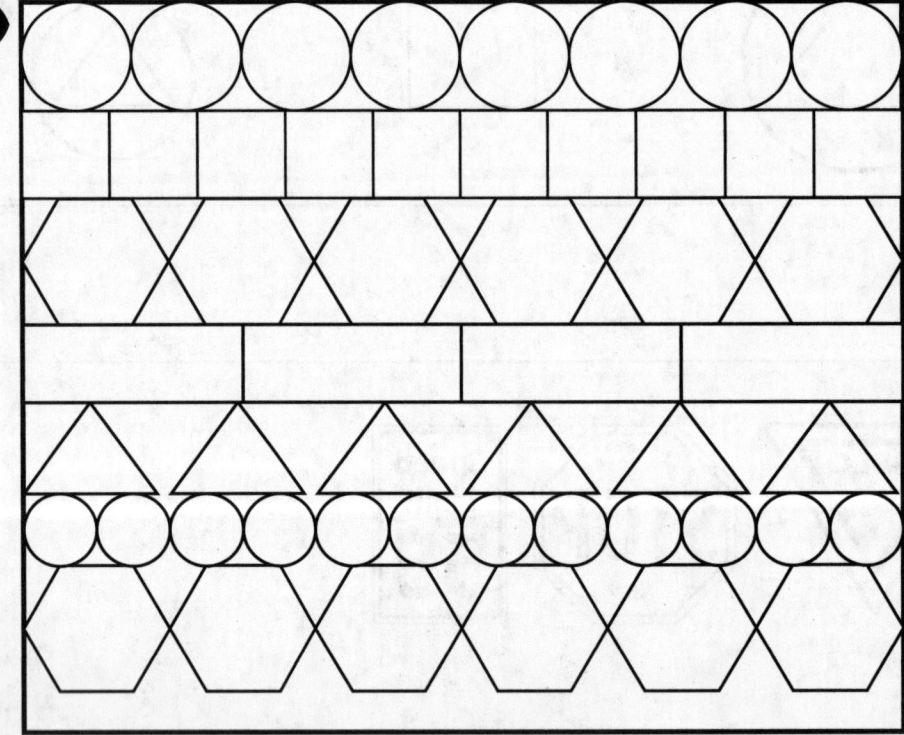

 **Math @ Home Activity**

While eating dinner, ask your child how many sides a triangle has. Then ask him or her to identify triangles in the room. Challenge your child to brainstorm different foods that are shaped like a triangle, such as a piece of pizza and a tortilla chip.

Directions: 3. Which shapes are triangles? Circle the triangles. **4.** Which shapes are triangles? Color the triangles.

Lesson 5-2

Additional Practice

Name _____

Review

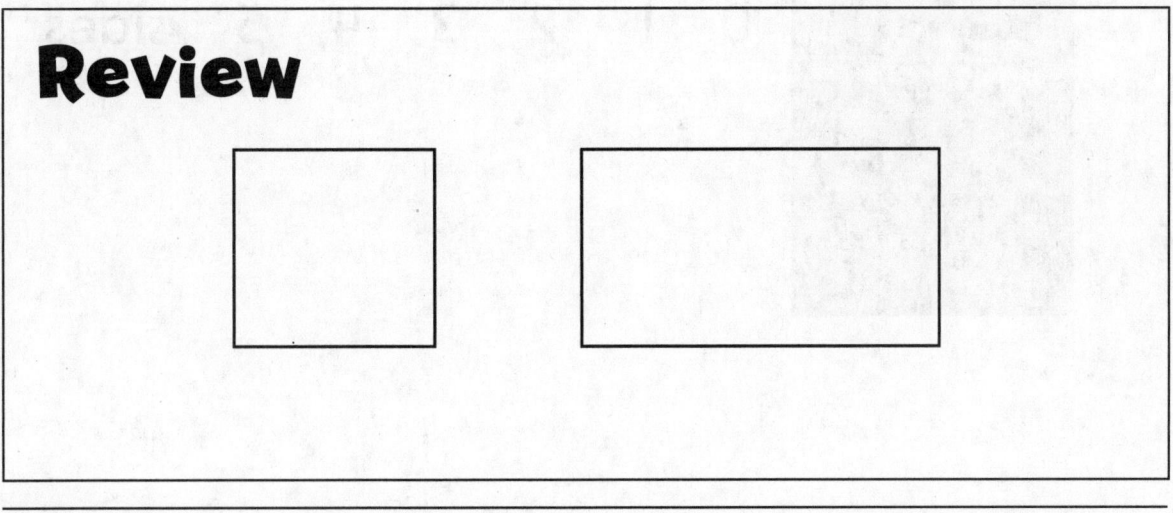

1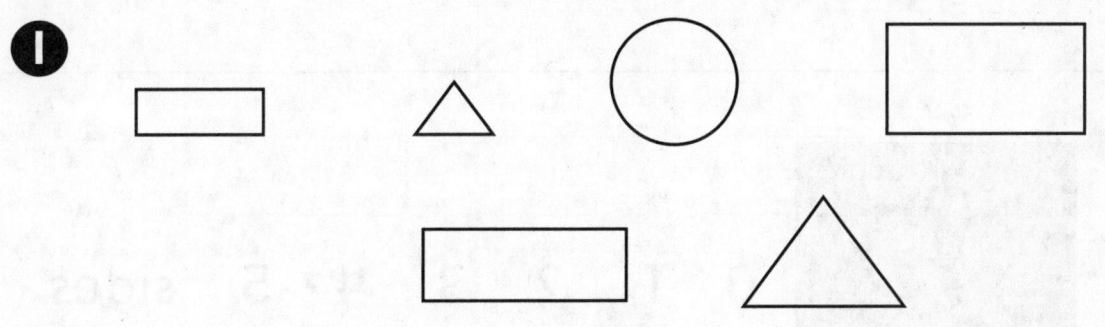

2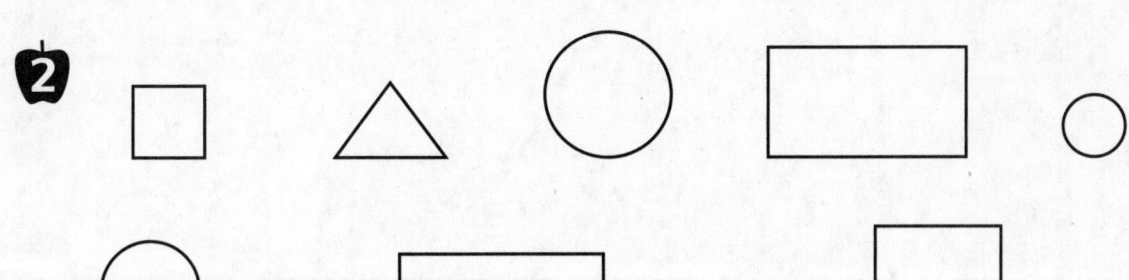

Review: You can identify and name a square and a rectangle. A square has sides that are all the same. A rectangle can have two short sides and two long sides.

Directions: 1. Which shapes are rectangles? Circle the rectangles. **2.** Which shapes are squares? Circle the squares.

0 1 2 3 4 5 sides

4

0 1 2 3 4 5 sides

Math @ Home Activity

Practice identifying squares and rectangles with your child. Play a game where you spot a rectangle in the room and ask your child to identify the object you spotted. Repeat with a square. Take turns spotting and identifying squares and rectangles.

Directions: 3–4. How many sides? Circle the number to show how many.

Lesson 5-3
Additional Practice

Name _____

Review

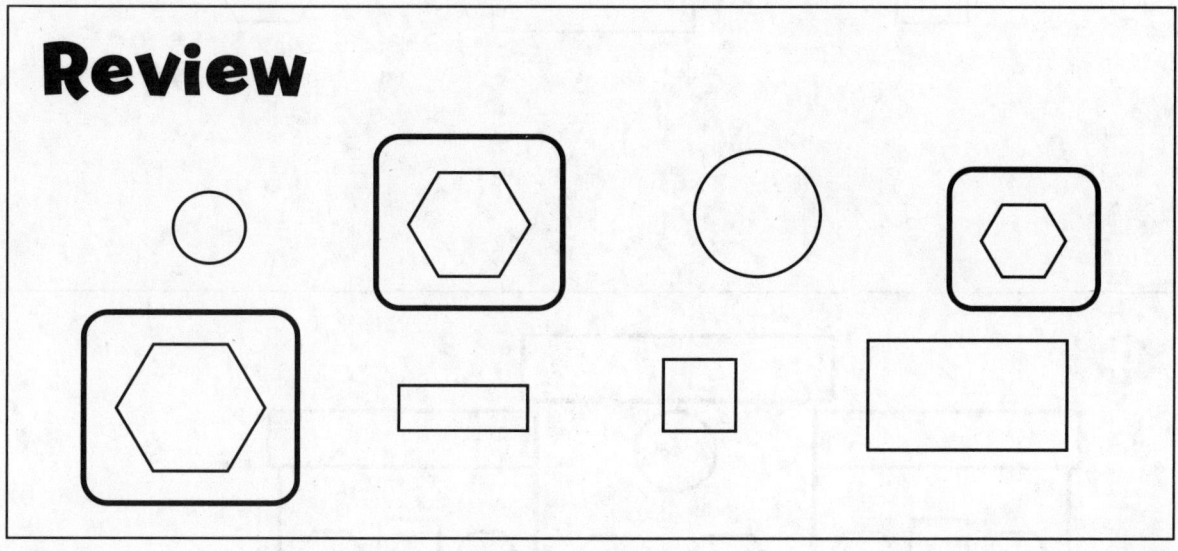

①

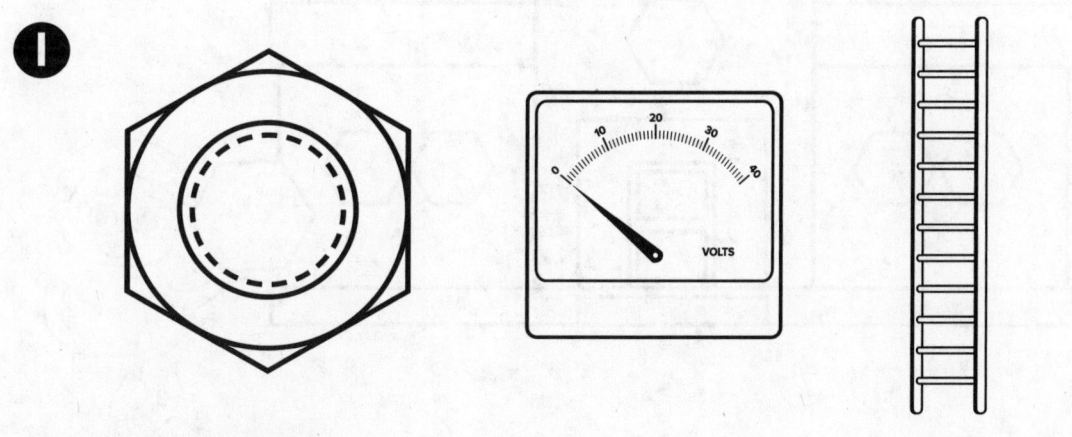

②

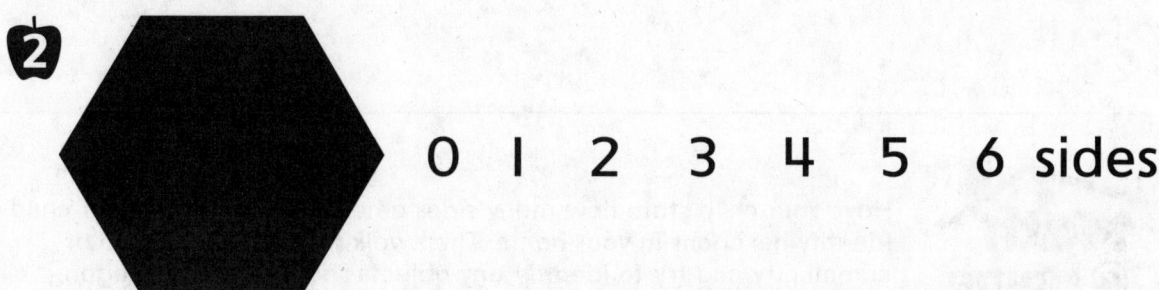

0 1 2 3 4 5 6 sides

Review: You can identify and name a hexagon. A hexagon has 6 sides.

Directions: 1. Which object is shaped like a hexagon? Circle the object that is shaped like a hexagon. **2.** How many sides? Circle the number to show how many.

3

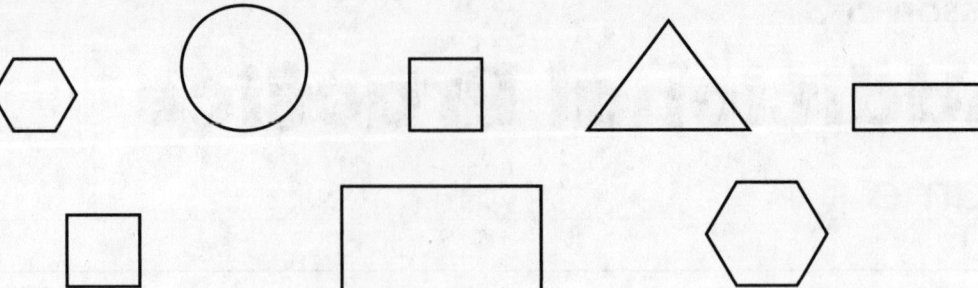

4

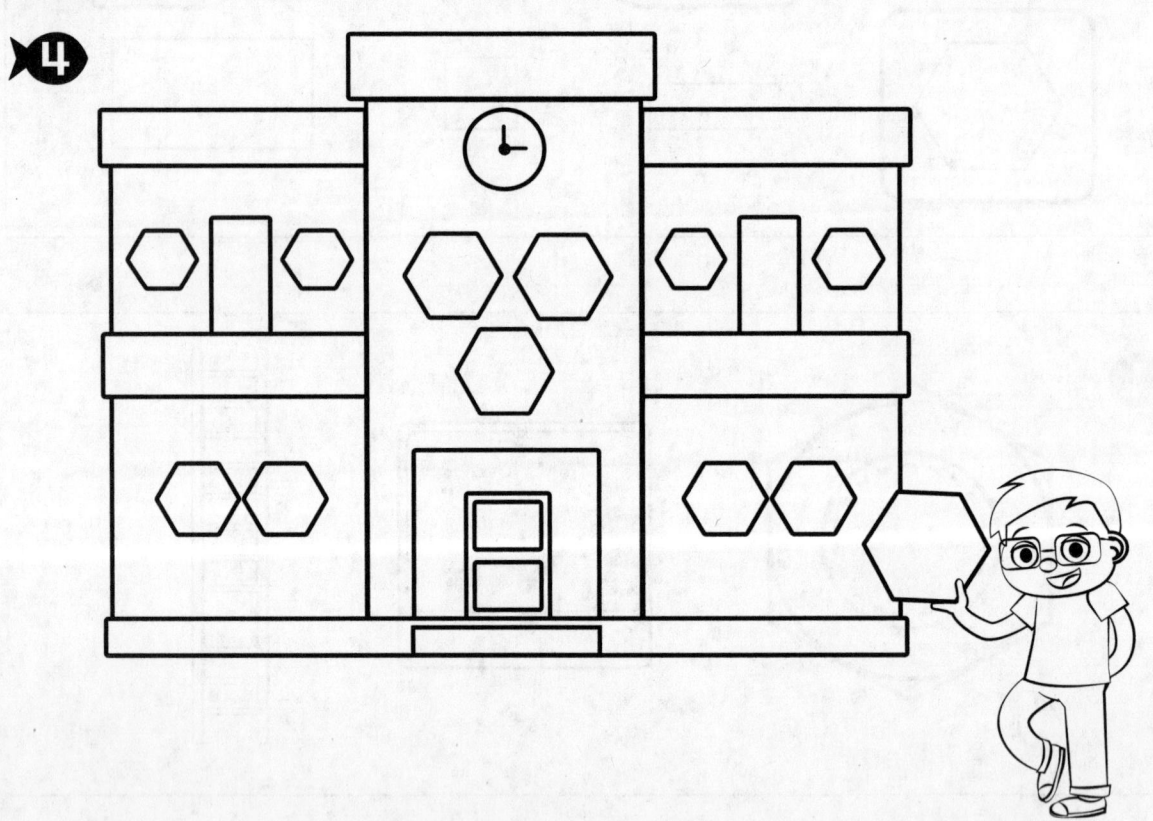

Math @Home Activity

Have your child state how many sides a hexagon has. With your child, identify hexagons in your home. Then walk through your home or community and try to identify any objects shaped like a hexagon, such as a bolt or the pattern on a soccer ball.

Directions: 3. Which shapes are hexagons? Circle the hexagons. **4.** Color the hexagons in the picture.

Lesson 5-4

Additional Practice

Name _____

Review

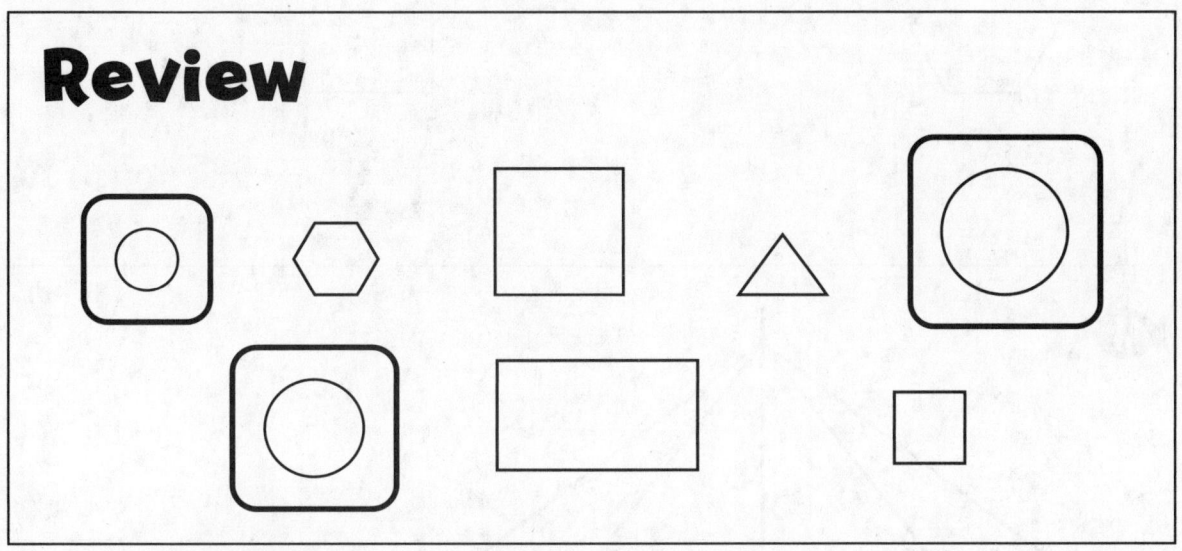

① 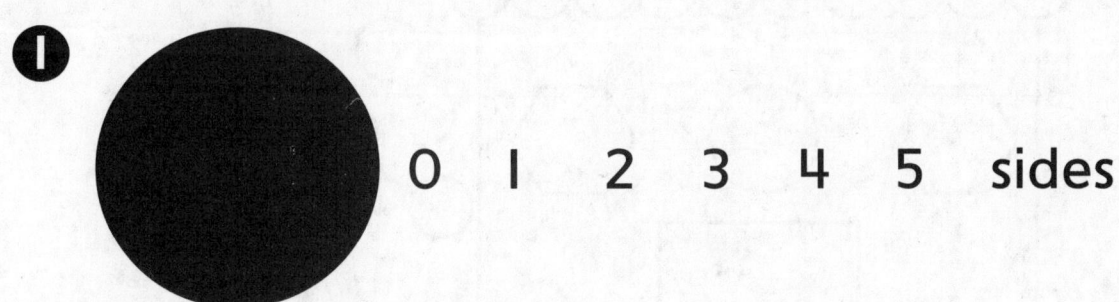   0   1   2   3   4   5   sides

②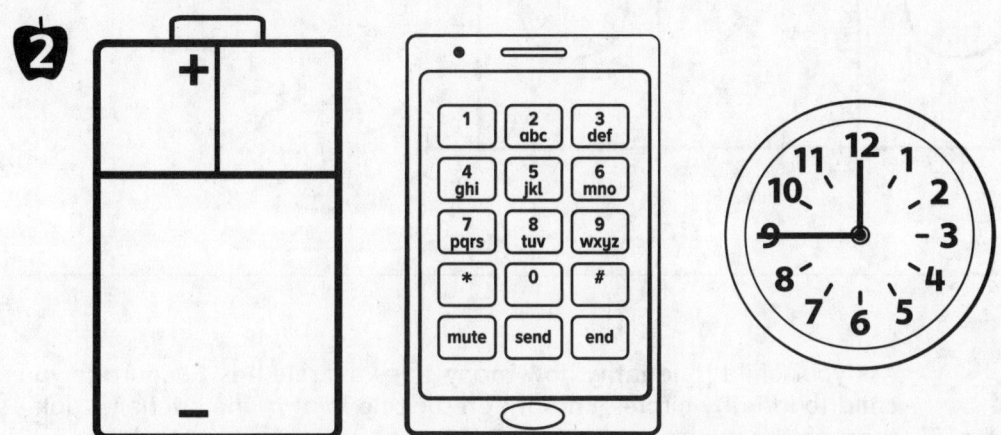

Review: You can identify and name a circle. A circle has 0 sides.

Directions: 1. How many sides? Circle the number to show how many. **2.** Which object is shaped like a circle? Circle the object that is shaped like a circle.

3

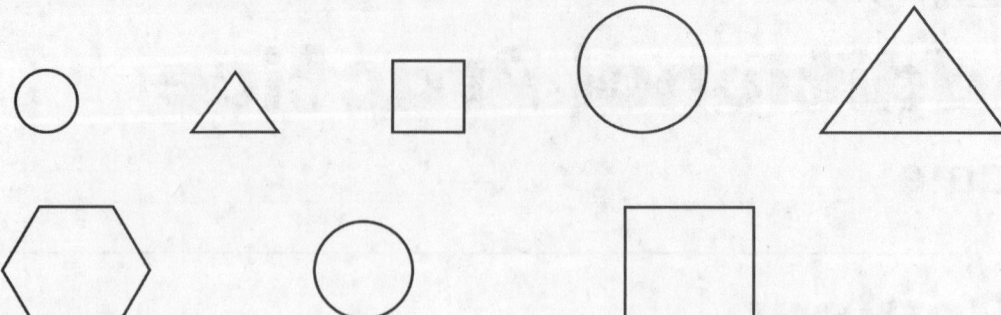

4

Math @ Home Activity

Ask your child to identify how many sides a circle has. Encourage your child to identify circles. Embark on a circle hunt in the kitchen. Look through the cupboards, refrigerator, etc., and ask your child to identify circles he or she sees.

Directions: 3. Which shapes are circles? Circle the circles. **4.** Color the circles in the picture.

Student Practice Book

Additional Practice

Name _____

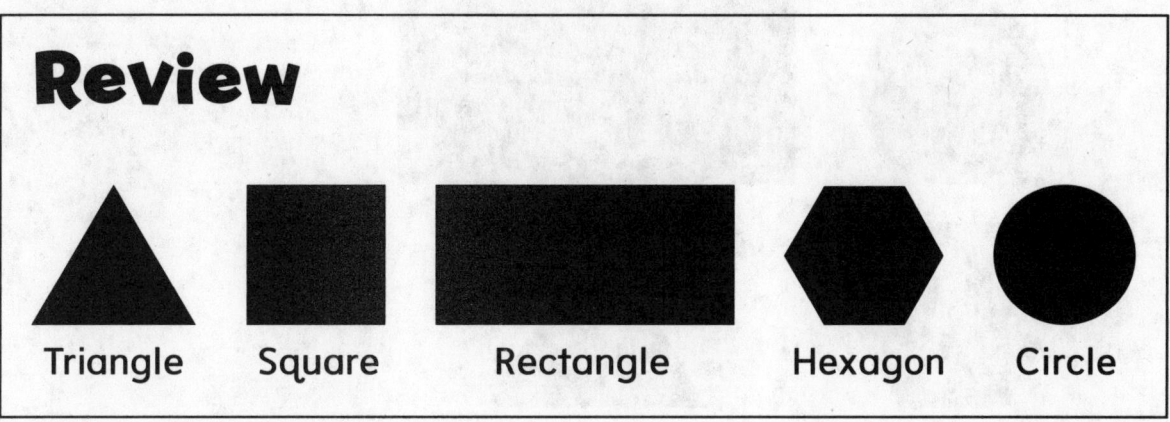

Review

Triangle   Square   Rectangle   Hexagon   Circle

1

2

Review: You can name shapes and describe their location. The square is next to the triangle.

Directions: 1. Which object is *beside* the beach ball? Circle the object and describe its shape. Which object is *below* the beach ball? Draw an X on the object and describe its shape. **2.** Which shape is *next to* the hexagon? Circle the shape and describe it. Which shape is *above* the hexagon? Draw an X on the shape and describe it.

3

4

Math @ Home Activity

Give your child many opportunities to identify and name 2-dimensional shapes as well as describe the location of each shape. For example, ask your child to identify an object that is shaped like a rectangle. Then ask your child to describe the location of the object. Repeat with other shapes.

Directions: 3. Which shape is *next to* the circle? Draw an X on the shape and describe it. Which shape is *above* the hexagon? Circle the shape and describe it. **4.** Which shape is *below* the hexagon? Draw an X on the shape and describe it. Which shape is *beside* the circle? Circle the shape and describe it.

Additional Practice

Name _____

Review

    3 and 2 is 5.

    1 and 3 is 4.

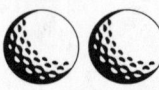 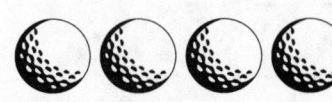   2 and 4 is 6.

❶

_____
_ _ _.

2 and 1 is _____.

❷

_____
_ _ _.

3 and 3 is _____.

Review: You can add groups of objects together to represent an addition story.

Directions: Use counters or drawings to represent each addition story. Write the total.
1. Grace makes 2 baskets in a row. She makes one more basket. How many baskets does Grace make in all? **2.** 3 children are playing volleyball. 3 more children join them. How many children are playing volleyball now?

3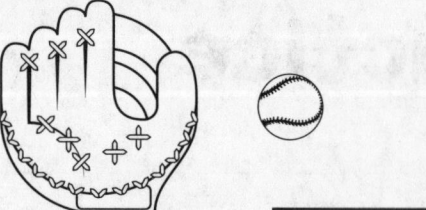

_____

_ _ _

4 and 3 is _____.

 **4**

_____

_ _ _

2 and 6 is _____.

Math @ Home Activity

Give your child many opportunities to practice putting together two parts to find the total. For example, while cleaning up toys, have your child separate the toys into two parts. Have your child count the toys in each part. Then ask him or her how many toys there are in all.

Directions: Write the total for each addition story. **3.** Manuel was playing catch with his sister. He caught four baseballs in a row. He caught three more. How many baseballs did he catch now? **4.** Two children are playing soccer. Six more children join them. How many children are playing soccer now?

Additional Practice

Name _____

Review

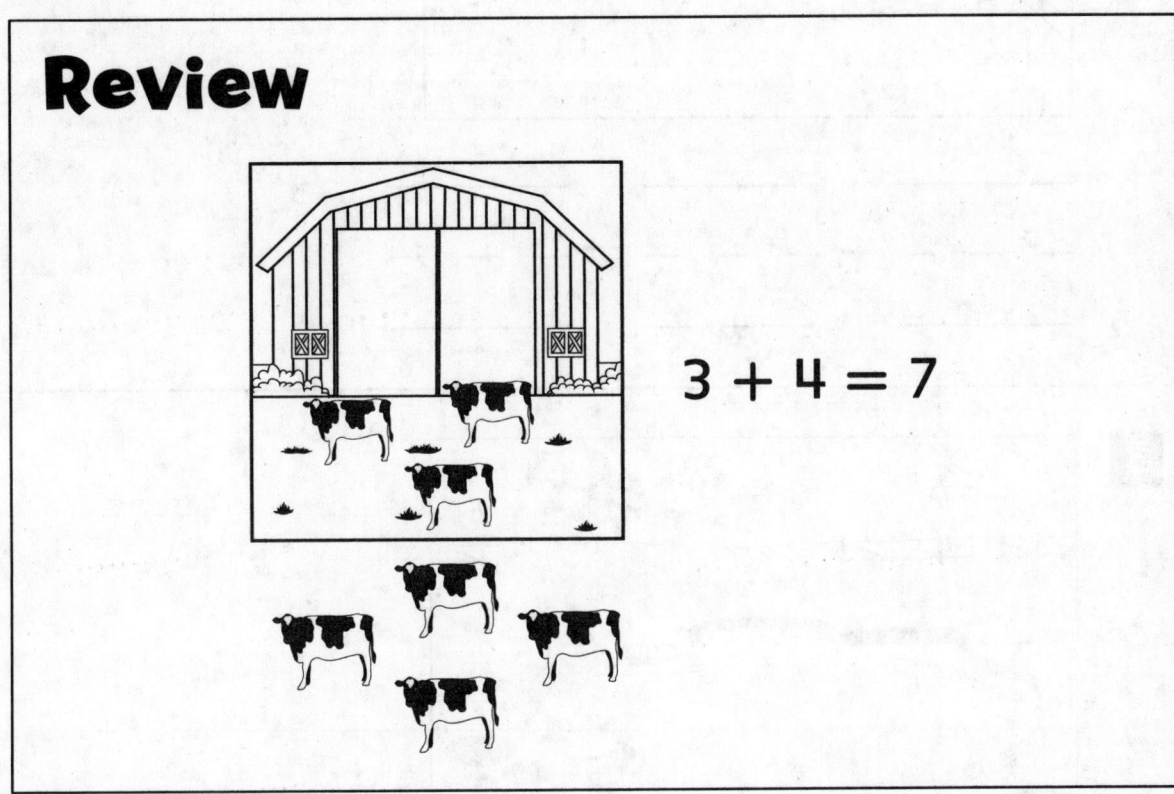

$3 + 4 = 7$

1

$3 + 5 = \underline{\ \ \ } \; \text{-} \; \text{-} \; \text{-}$

Review: You can represent a word problem with drawings. Add the parts to find the total. There are 3 cows in the barnyard. 4 more cows walk over. How many cows are in the barnyard now?

Directions: 1. There are 3 chickens on the farm, 5 more chickens walk over. How many chickens are on the farm now? Represent the word problem with drawings. Add to find the total. Write the total.

2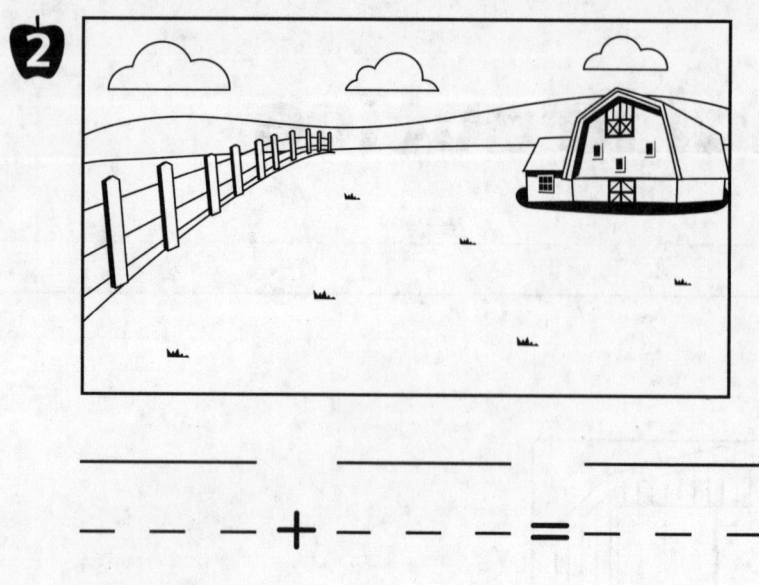

___ ___ ___   +   ___ ___ ___   =   ___ ___ ___

___ ___ ___   ___ ___ ___   ___ ___ ___

3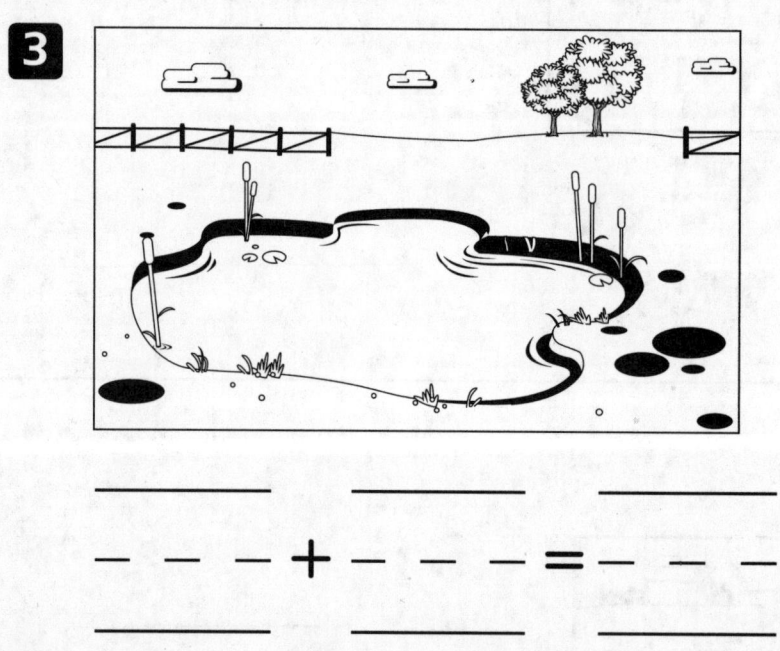

___ ___ ___   +   ___ ___ ___   =   ___ ___ ___

___ ___ ___   ___ ___ ___   ___ ___ ___

Math @ Home Activity

While reading a story, ask your child to determine the number of objects on the page if a certain number of the same object are added. Practice adding on each page of the book.

Directions: 2. There is I dog in the field. I more dog runs over. How many dogs are in the field now? Represent the word problem with drawings. Write an equation to match.
3. There are 7 beavers in the water. 0 more beavers swim over. How many beavers are in the water now? Represent the word problem with drawings. Write an equation to match.

Additional Practice

Name _____

Review

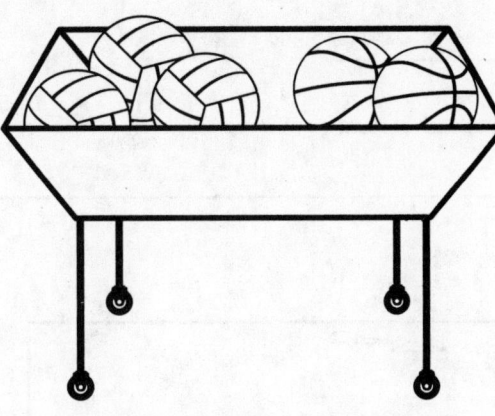

$$3 + 2 = 5$$

①

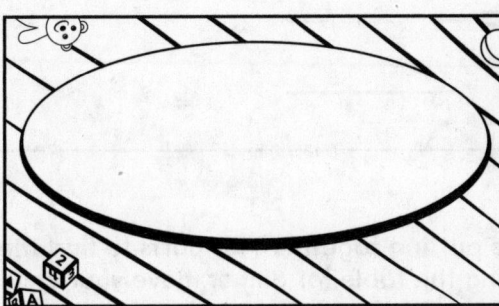

$$4 + 2 = \underline{\quad}$$

_____

Review: Coach Matlock puts 3 volleyballs and 2 basketballs in the ball cart. How many sports balls are in the cart? You can represent a word problem with counters or drawings to find the total.

Directions: I. The librarian puts 4 big books and 2 small books on the rug. How many books are on the rug? Use counters or drawings to represent the addition story. Then write the total.

Student Practice Book

2

$$5 + 3 = \overline{}$$
$$\overline{}$$

3

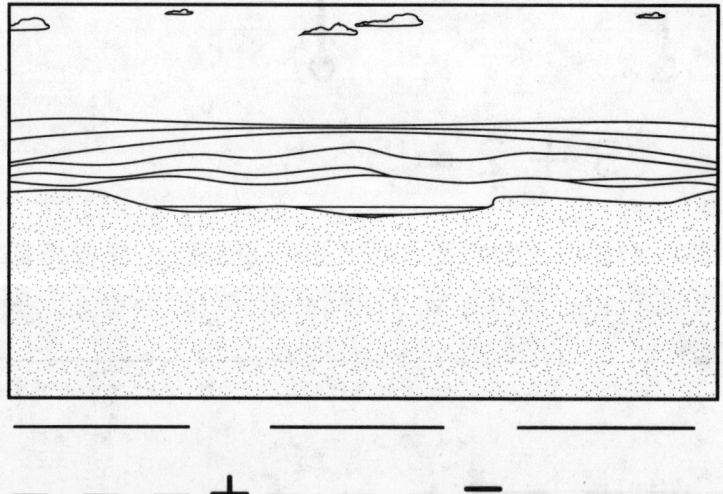

$$\underline{} + \underline{} = \underline{}$$

Copyright © McGraw-Hill Education

Math @ Home Activity

Have your child practice putting together two parts to find the total. For example, while setting the table for dinner, have your child represent the total number of cups, plates, or chairs with an addition sentence.

Directions: 2. 5 dogs and 3 cats are playing in the yard. How many animals are in the yard? Use counters or drawings to represent the addition story. Then write the total.
3. There are 7 beach balls and 2 shells on the beach. How many objects are on the beach? Use counters or drawings to represent the addition story. Then write an equation.

Additional Practice

Name _____

Review

$$4 + 3 = 7$$

1

_____ _____ _____

- - - - - - - - -

_____ _____ _____

_____

2

_____ _____ _____

- - - - - - - - -

_____ _____ _____

Review: There are 4 toy whales on the floor and 3 toy whales in the toy box. How many toy whales are there in all? You can read and represent the word problem. You can also write an equation to match your thinking.

Directions: 1. Darby buys 9 puppets. Some are horses and some are goats. How many of each puppet does she buy? Use counters or drawings to show one way to solve the problem. Then write an equation. **2.** Robyn has 5 keys. Her mom gives her one more key. How many keys does Robyn have in all? Robyn says there are 4 keys. Use counters or drawings to help correct her thinking. Then write an equation.

3

_____   _____   _____

- - - - -   - - - - -   - - - - -

_____   _____   _____

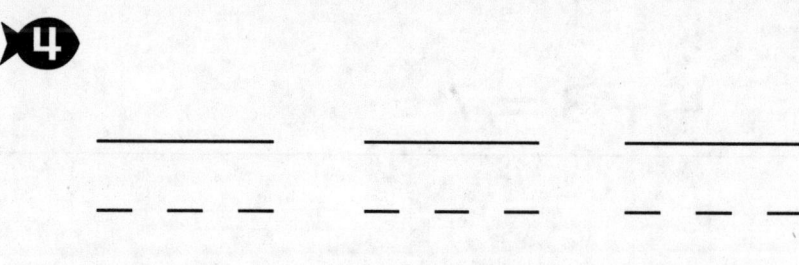 **4**

_____   _____   _____

- - - - -   - - - - -   - - - - -

_____   _____   _____

 **Math @ Home Activity**

While outside, create a word problem for your child to solve. Have your child find objects to represent the problem. Then have your child state an equation that matches the representation. Repeat, switching roles so that your child creates a word problem that you solve using a representation.

Directions: 3. Siu has 6 toy rings. She gives some to her sister and some to her brother. Use counters or drawings to show one way to solve the problem. Then write an equation.
4. Ahmad uses 4 buckets to make a sand castle. Then he uses 4 more buckets. How many buckets does Ahmad use in all? Ahmad says there are 9 buckets of sand. Use counters or drawings to help correct his thinking. Then write an equation.

Student Practice Book

Additional Practice

Name _____

Review

● + ● ● ● ● $5 = 1 + 4$
● ● + ● ● ● $5 = 2 + 3$
● ● ● + ● ● $5 = 3 + 2$
● ● ● ● + ● $5 = 4 + 1$

1

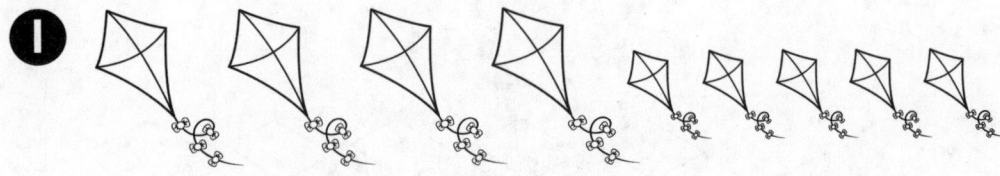

$$9 = \underline{} \underline{} \underline{} + \underline{} \underline{} \underline{}$$

2

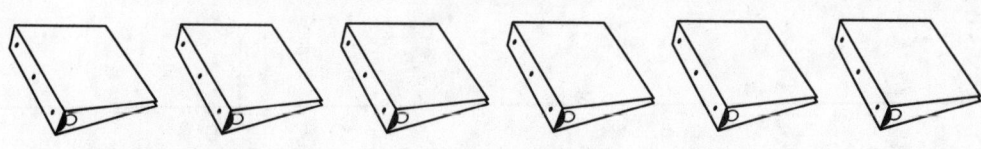

$$6 = \underline{} \underline{} \underline{} + \underline{} \underline{} \underline{}$$

Review: There are 5 apples. Mana and Elke each receive some apples. You can represent the word problem with counters to show all the possible ways to find a total of 5.

Directions: 1. Write the missing addends to match the representation. **2.** Alba buys 6 books. Some are red and some are green. How many of each color could she have? Color to show one way. Then write an equation to match.

3

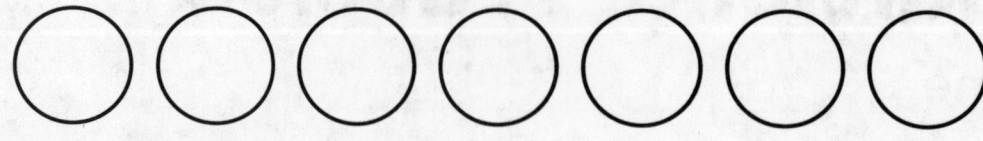

$$7 = \underline{} \; \overline{} + \underline{} \; \overline{}$$

4

$$8 = \underline{} \; \overline{} + \underline{} \; \overline{}$$

Math @ Home Activity

Give your child 10 or fewer small objects. Have your child separate the objects to show two addends that equal the total number of objects. Then have your child separate the objects in different ways to show other addends that equal the total number of objects.

Directions: 3. Myong has baseballs and tennis balls. She has 7 sports balls in all. How many of each ball can she have? Draw to show one way. Then write an equation to match.
4. There are 8 books in a basket. Some of the books are about dolls and some are about cars. How many books could be about dolls and how many could be about cars? Draw to show one way. Write the missing addends to match your representation. Then write an equation to match.

Additional Practice

Name _____

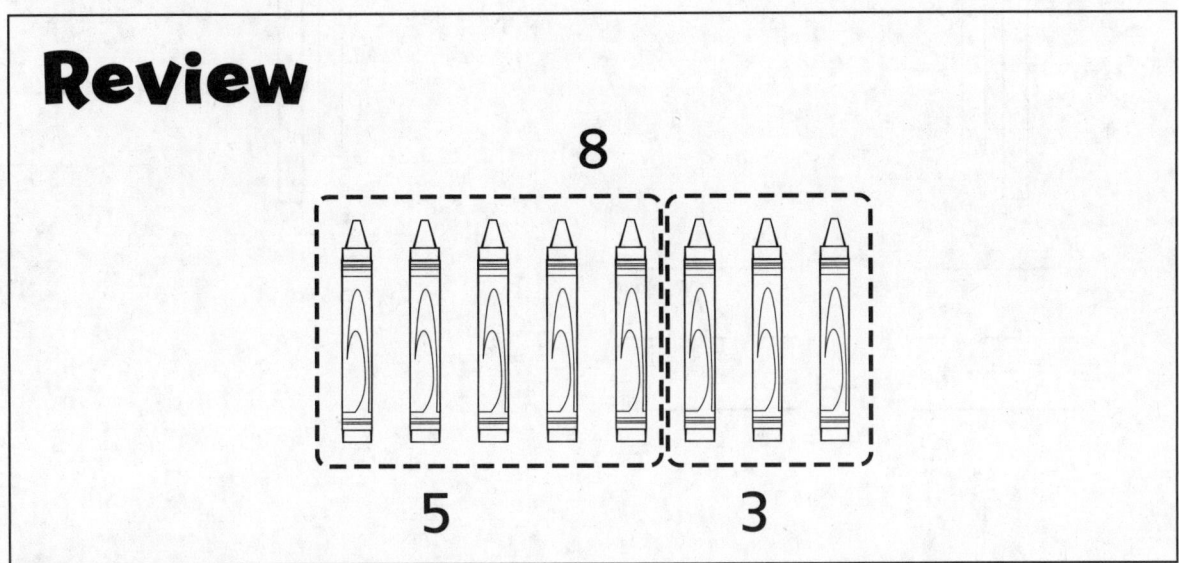

Review

8

5   3

①

_____

_ _ _ _ _ _ _ _ _ _

_____ and _____

②

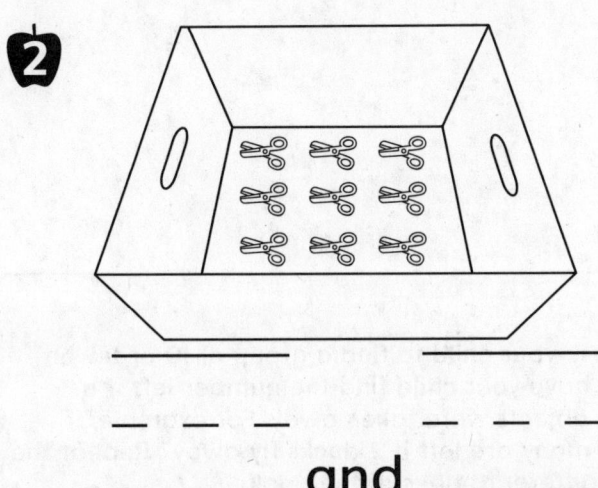

_____

_ _ _ _ _ _ _ _ _ _

_____ and _____

Review: You can take apart a group by drawing circles around the two parts.

Directions: 1. How can you take apart the group of books? Draw circles and write numbers to show the parts. **2.** How can you take apart the group of scissors? Draw circles and write numbers to show the parts.

Student Practice Book

3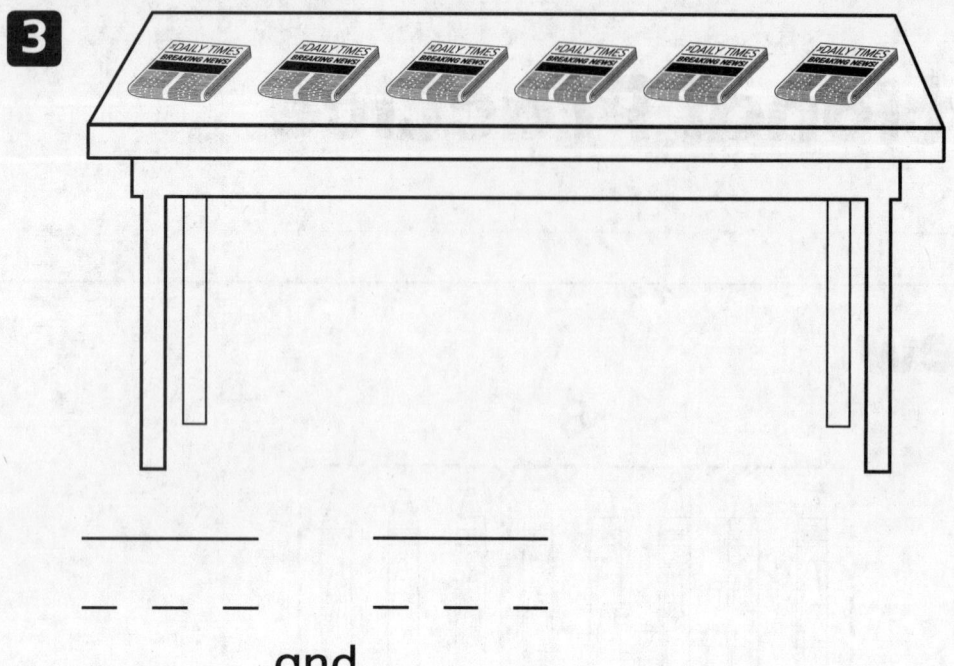

_____ _____

_____ and _____

4

_____ _____

_____ and _____

Math @ Home Activity

While reading a book, ask your child to find a group of 10 or fewer objects on a page. Then have your child find the number left if a certain number of those objects were taken away. For example, if 5 ducks are shown, how many are left if 2 ducks fly away? Repeat the activity with objects on different pages of the book.

Directions: 3. How can you take apart the group of newspapers? Draw circles and write numbers to show the parts. **4.** How can you take apart the group of bones? Draw circles and write numbers to show the parts.

Additional Practice

Name _____

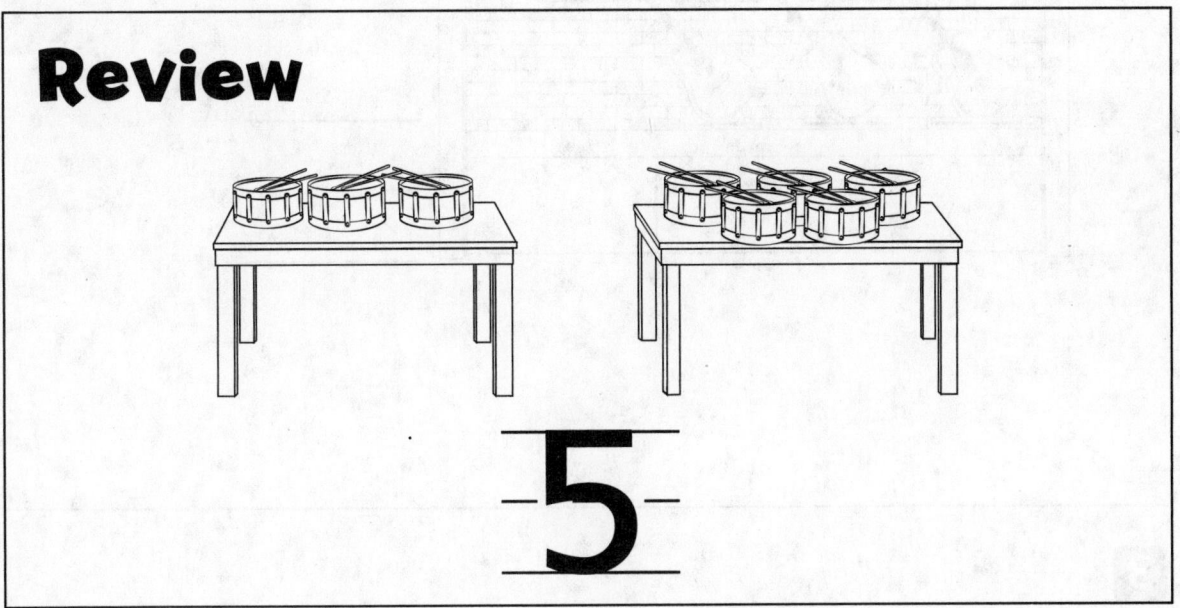

Review

-5

①

Review: You can represent a story problem with drawings. The band director has 8 drums. He puts 3 drums on one table, and the rest on another table. You can write the difference.

Directions: 1. A music teacher has 9 sheets of music. She gives out 6 sheets. How many sheets of music does she have left? Use counters to model the subtraction story. Write the difference.

2

_____

_ _ _ _ _ _ _

_____

3

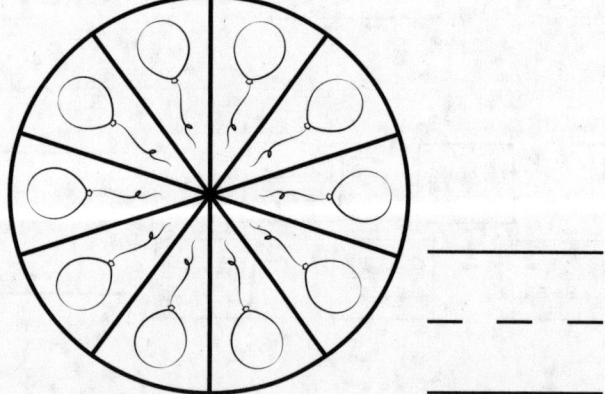

_____

_ _ _ _ _ _ _

_____

Math @ Home Activity

Write a number between 1 and 10. Have your child draw a picture to show taking the number apart. Then have your child write an equation for the representation. Challenge your child to take apart the number in as many different ways as possible. For example, the equations $3 - 0 = 3$, $3 - 1 = 2$, $3 - 2 = 1$, and $3 - 3 = 0$ can be written when taking apart 3.

Directions: 2. There are 6 guitars at the music store. 5 guitars are sold. How many guitars are left? Use counters to model the subtraction story. Write the difference. **3.** There are 10 balloons on the wall. Reynaldo popped 7 balloons. How many are left? Use counters to model the subtraction story. Write the difference.

Additional Practice

Name _____

Review

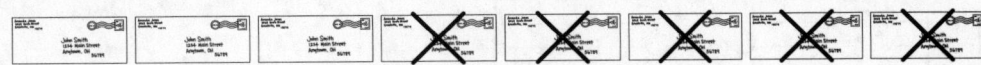

$$8 - 5 = 3$$

1

___ ___ ___ ___ ___ = ___ ___ ___

2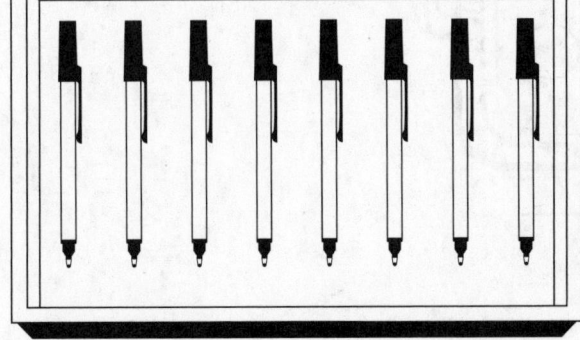

___ ___ ___ ___ ___ = ___ ___ ___

Review: John receives 8 letters in the mail. 5 letters are from Amanda. How many letters are not from Amanda? You can represent a story problem with drawings. You can write an equation to match.

Directions: 1–2. Write numbers to complete the equations. **1.** A store has 10 mail boxes. They sell 6 mail boxes. How many are left? **2.** Carlos has 8 pens. He gives 2 pens away. How many are left?

3

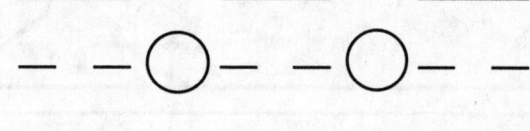

4

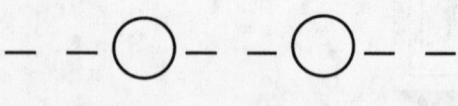

Math @ Home Activity

Give your child many opportunities to solve subtraction word problems. Place 8 apple slices on a plate. Take away 6 slices and set them on another plate. Have your child write and solve a subtraction equation to represent the situation. Repeat the activity using different numbers.

Directions: 3. Shyla has 7 stickers on her paper. She loses 6 stickers. How many are left? Write an equation to match. **4.** Dionne has 9 balls of yarn. He uses 5 balls of yarn. How many are left? Write an equation to match.

Additional Practice

Name _____

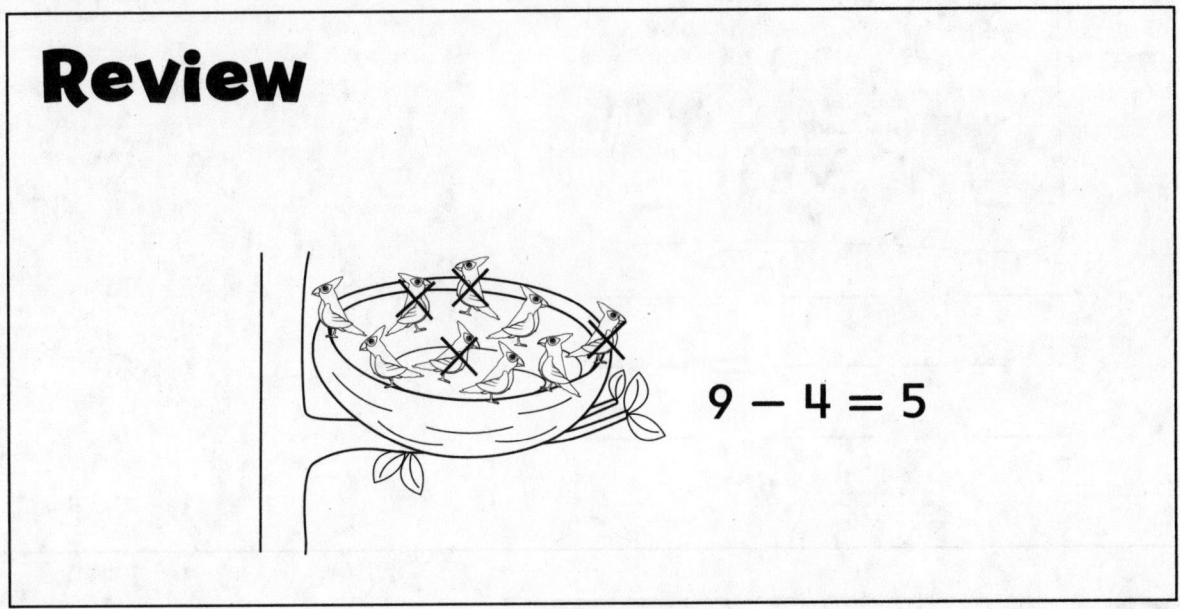

Review

$$9 - 4 = 5$$

1

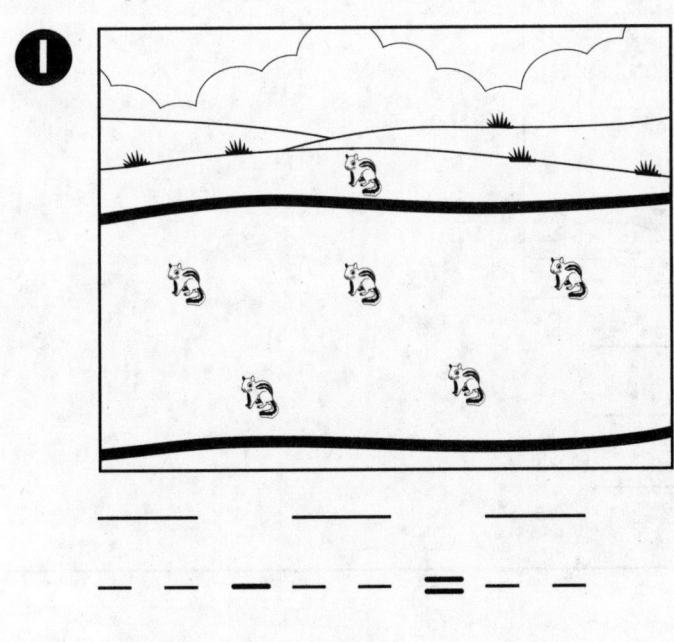

‾‾‾‾ ‾‾‾‾ ‾‾‾‾ ‾‾‾‾

‾‾‾‾ ‾ ‾ ‾ ‾ = ‾ ‾

‾‾‾‾ ‾‾‾‾ ‾‾‾‾

Review: 9 birds are in a nest. 4 birds fly away. How many are left in the nest? You can represent a story problem with drawings. You can write an equation to match.

Directions: 1. Hilma sees 6 chipmunks on the ground. 5 chipmunks climb a tree. How many are left on the ground? Solve. Write numbers and trace symbols to complete the equation.

Student Practice Book
77

2

___   ___   ___

___ − ___ = ___

___   ___   ___

3

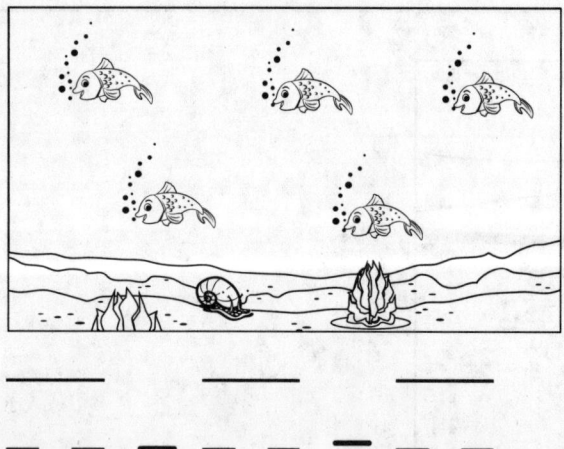

___   ___   ___

___ − ___ = ___

___   ___   ___

Math @ Home Activity

Create a word problem related to a situation around your home. Ask your child to find objects to represent and "act out" the problem. Then your child can write an equation to match.

Directions: 2. 8 cows are eating grass in a field. 6 cows lie down. How many are still standing? Solve. Write numbers and trace symbols to complete the equation. **3.** 5 fish are near the bottom of the pond. 5 fish swim away. How many are left? Solve. Write numbers and trace symbols to complete the equation.

Additional Practice

Name ....

Review

$$4 + 5 = 9$$

 1

_____ _____ _____

\- \- \- ◯ \- \- \- ◯ \- \- \-

_____ _____ _____

 2

_____ _____ _____

\- \- \- ◯ \- \- \- ◯ \- \- \-

_____ _____ _____

Review: Phuong plants 4 flowers in his garden. The next day he plants 5 more flowers. How many flowers are there now? You can represent a story problem with drawings. You can write an equation to match.

Directions: 1–2. Draw to represent the problem. Then write an equation to match. **1.** Monroe puts 2 eggs in his basket. Then he adds 5 more eggs. How many eggs are in his basket now? **2.** 6 bees land on a flower. 2 more bees land. How many bees are there now?

3

_____ _____ _____

- - - ◯ - - - ◯ - - -

_____ _____ _____

4

_____ _____ _____

- - - ◯ - - - ◯ - - -

_____ _____ _____

Math
@ Home
Activity

Gather 9 small objects. Ask your child how many would be left if you took away 8 of the objects. Have your child write an equation to represent the situation. Repeat the activity with different numbers of objects.

Directions: 3. There are 8 trees in a field. A farmer cuts down 8 trees. How many trees are there now? Solve. Write an equation to match. **4.** There are 7 flowers. Ilene picks 5 flowers. How many are left? Solve. Write an equation to match.

Student Practice Book
80

Additional Practice

Name _____

Review

$$2 + 2 = 4$$

---

❶ $2 + 1 =$ _____

Review: You can use a number path to help you add.

Directions: 1. How can you count on to find $2 + 1$? Color the starting number. Count on 1. Circle the stopping number. What is the sum?

2 3 + 2 = — — —

3 1 + 3 = — — —

| 1 | 2 | 3 | 4 | 5 |

4 2 + 3 = — — —

| 1 | 2 | 3 | 4 | 5 |

Math @ Home Activity

Work with your child to develop fluency with adding within 5. For example, have your child use shapes cut from paper to show a number within 5. Then have your child rearrange the shapes to show another way to show that number. Repeat with other numbers within 5.

Directions: 2. How can you count on to find 3 + 2? Color the starting number. Count on 2. Circle the stopping number. What is the sum? **3.** How can you count on to find 1 + 3? Color the starting number. Count on 3. Circle the stopping number. What is the sum? **4.** How can you count on to find 2 + 3? Color the starting number. Count on 3. Circle the stopping number. What is the sum?

Additional Practice

Name _____

Review

$$5 - 2 = 3$$

---

❶ $4 - 2 =$ _ _ _ _

Review: You can use a number path to help you subtract by counting back.

Directions: I. How can you count back to find 4 − 2? Color the starting number. Count back 2. Circle the stopping number. What is the difference?

Student Practice Book

83

2 $3 - 2 =$ ___ ___

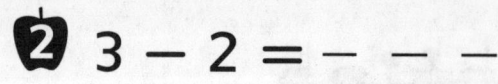

| 1 | 2 | 3 | 4 | 5 |

3 $4 - 3 =$ ___ ___

| 1 | 2 | 3 | 4 | 5 |

4 $5 - 2 =$ ___ ___

| 1 | 2 | 3 | 4 | 5 |

Math @ Home Activity

Work with your child to practice breaking apart groups within 5. For example, have your child use building blocks to show 5. Then have your child arrange the blocks to show different ways to break apart 5. Repeat the activity with other numbers within 5.

Directions: 2. How can you count back to find 3 − 2? Color the starting number. Count back 2. Circle the stopping number. What is the difference? **3.** How can you count back to find 4 − 3? Color the starting number. Count back 3. Circle the stopping number. What is the difference? **4.** How can you count back to find 5 − 2? Color the starting number. Count back 2. Circle the stopping number. What is the difference?

Additional Practice

Name _____

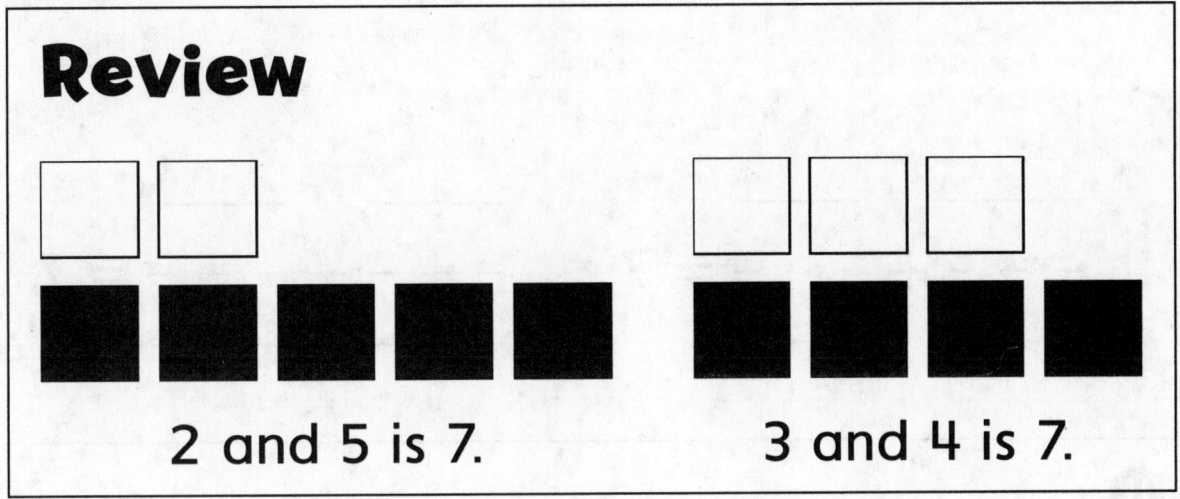

Review

2 and 5 is 7.   3 and 4 is 7.

1

$3 + 3 = 6$

2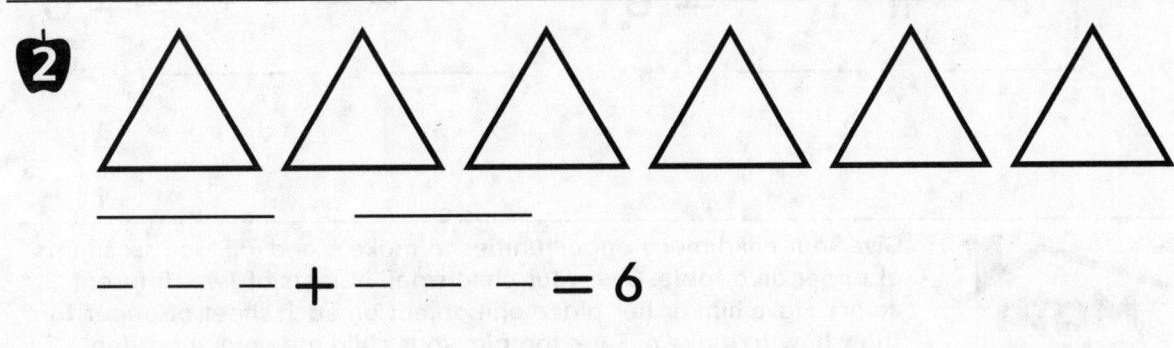

___ ___

___ ___ ___ + ___ ___ ___ = 6

___ ___

Review: You can show ways to make 6 and 7. The squares show two different ways to make 7.

Directions: 1. How can you make 6? Color the triangles using pink and brown to show 3 and 3 is 6. **2.** How can you make 6? Use two colors to show a way to make 6. Write the equation to match.

③

$$\underline{\qquad}\ \underline{\qquad} \qquad \underline{\qquad}\ \underline{\qquad}$$
$$\underline{\ \ \ } + \underline{\ \ \ } = 7 \qquad \underline{\ \ \ } + \underline{\ \ \ } = 7$$
$$\underline{\qquad}\ \underline{\qquad} \qquad \underline{\qquad}\ \underline{\qquad}$$

④

$$\underline{\qquad}\ \underline{\qquad} \qquad \underline{\qquad}\ \underline{\qquad}$$
$$\underline{\ \ \ } + \underline{\ \ \ } = 6 \qquad \underline{\ \ \ } + \underline{\ \ \ } = 6$$
$$\underline{\qquad}\ \underline{\qquad} \qquad \underline{\qquad}\ \underline{\qquad}$$

Math @ Home Activity

Give your child many opportunities to make 6 and 7. Place six sheets of paper on a table. Give your child small objects of two different colors. Have him or her place one object on each sheet of paper to show how to make 6. For example, your child may place I green marble on each of 4 sheets of paper and I blue marble on each of 2 sheets of paper to show that 4 and 2 make 6. Repeat with ways to make 7.

Directions: 3. How can you make 7? Draw counters to show two ways to make 7. Write the equations to match. **4.** How can you make 6? Draw counters to show two ways to make 6. Write the equations to match.

Additional Practice

Name

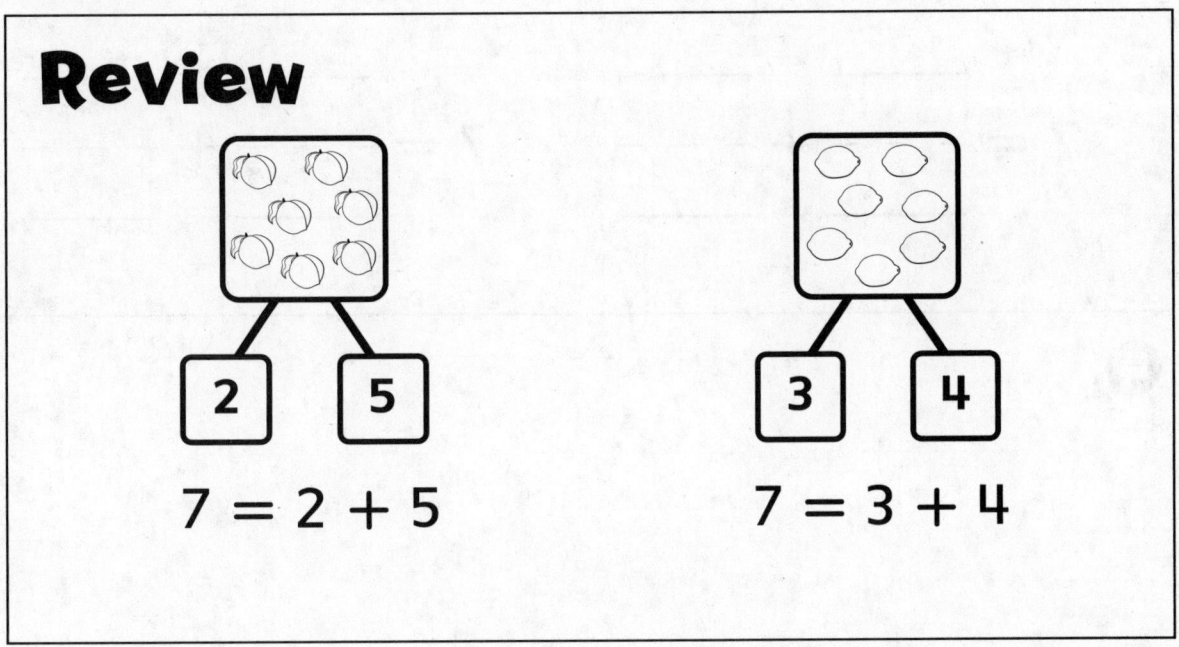

Review

$7 = 2 + 5$

$7 = 3 + 4$

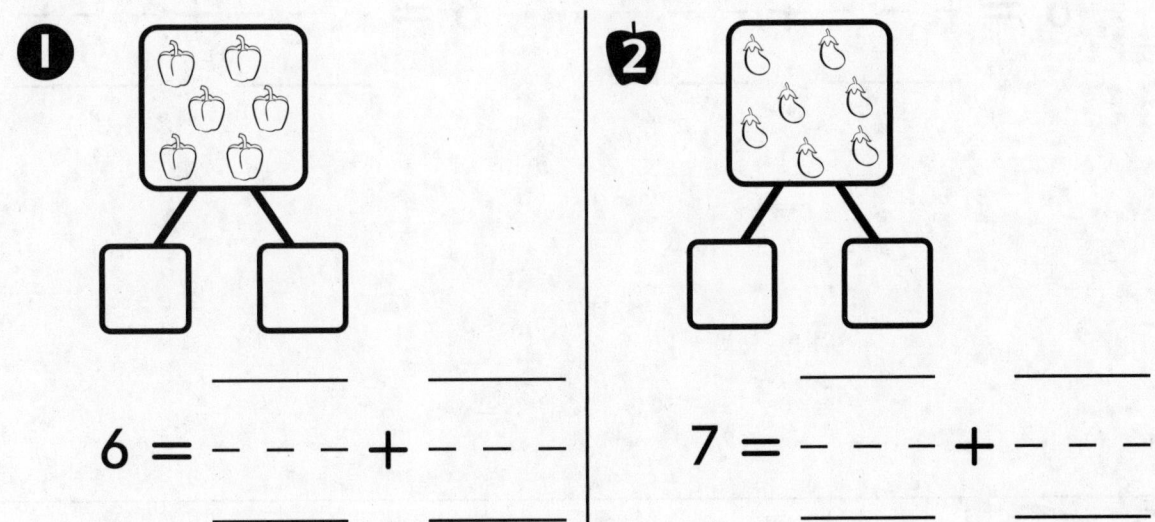

①

②

$6 = \text{---} + \text{---}$

$7 = \text{---} + \text{---}$

Review: You can break apart 6 and 7. The example shows two different ways to break apart 7.

Directions: 1. How can you decompose 6? Circle groups to show a way to decompose 6. Complete the number bond and the equation to match. **2.** How can you decompose 7? Circle groups to show a way to decompose 7. Complete the number bond and the equation to match.

❸

$$7 = \underline{} \; \overline{\underline{}} + \overline{\underline{}} \; \underline{} \qquad 7 = \underline{} \; \overline{\underline{}} + \overline{\underline{}} \; \underline{}$$

❹

$$6 = \underline{} \; \overline{\underline{}} + \overline{\underline{}} \; \underline{} \qquad 6 = \underline{} \; \overline{\underline{}} + \overline{\underline{}} \; \underline{}$$

Math @ Home Activity

Create a number bond, as seen in these exercises, using small containers or baskets in place of the three squares. Have your child use objects to show different ways to break apart 6 and 7, recording the numbers used each time. See whether your child can find all the ways to break apart the two numbers.

Directions: 3. How can you decompose 7? Draw connecting cubes to show two different ways to break apart 7. Write the equations to match. **4.** How can you decompose 6? Draw connecting cubes to show two different ways to break apart 6. Write the equations to match.

Additional Practice

Name _____

Review

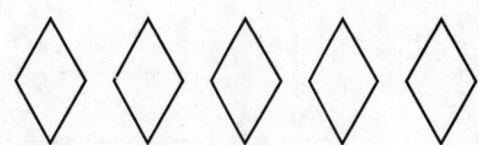

4 + 4 = 8   5 + 3 = 8

1

___ ___ ___ + ___ ___ ___ = 9

___ ___ ___   ___ ___ ___

2

___ ___ ___ + ___ ___ ___ = 8

___ ___ ___   ___ ___ ___

Review: You can show ways to make 8 and 9. This shows two different ways to make 8.

Directions: 1. How can you make 9? Use two colors to show one way to make 9. Write the equation to match. **2.** How can you make 8? Use two colors to show one way to make 8. Write the equation to match.

❸

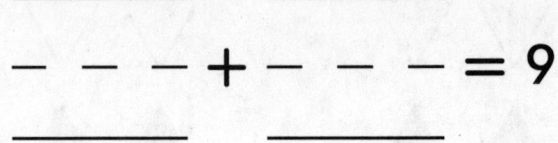

$$\underline{\quad\quad} \ \underline{\quad\quad}$$
$$\text{-} \ \text{-} \ \text{-} + \text{-} \ \text{-} \ \text{-} = 9$$
$$\underline{\quad\quad} \ \underline{\quad\quad}$$

❹

$$\underline{\quad\quad} \ \underline{\quad\quad}$$
$$\text{-} \ \text{-} \ \text{-} + \text{-} \ \text{-} \ \text{-} = 8$$
$$\underline{\quad\quad} \ \underline{\quad\quad}$$

Math @ Home Activity

Fold a sheet of paper in half vertically. Place 8 paperclips on the paper. Have your child move the paperclips to either side of the sheet of paper to show different ways to make 8. Place another paperclip on the paper, and have your child repeat the activity to show different ways to make 9.

Directions: 3. Draw 9 fish. Color to show one way to make 9. Write the equation to match. **4.** Draw 8 seashells. Color to show one way to make 8. Write the equation to match.

Additional Practice

Name _____

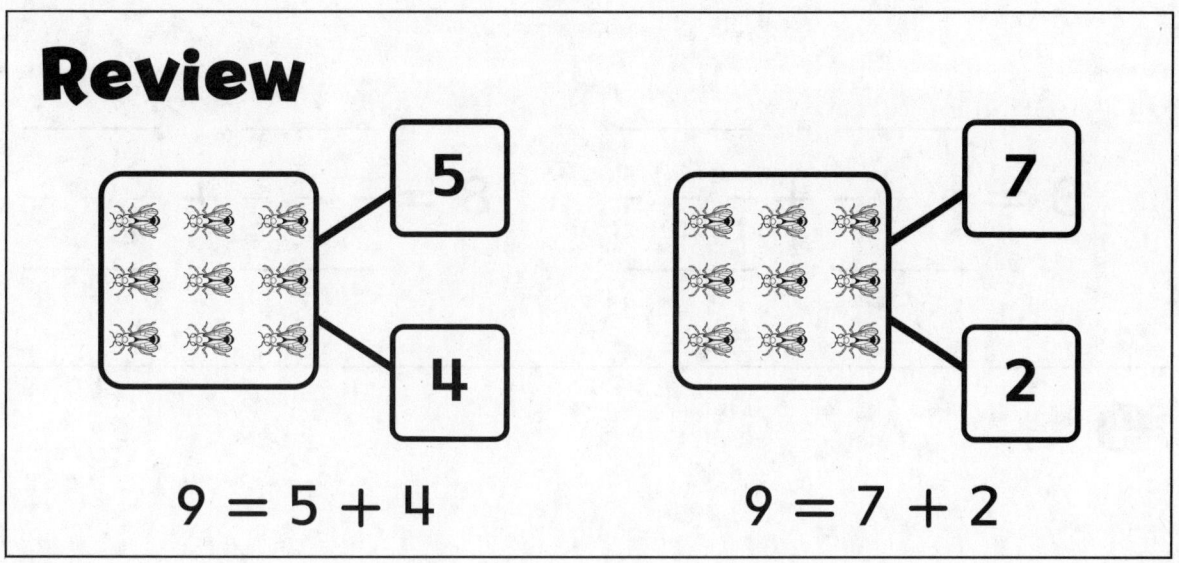

Review

$9 = 5 + 4$ $9 = 7 + 2$

1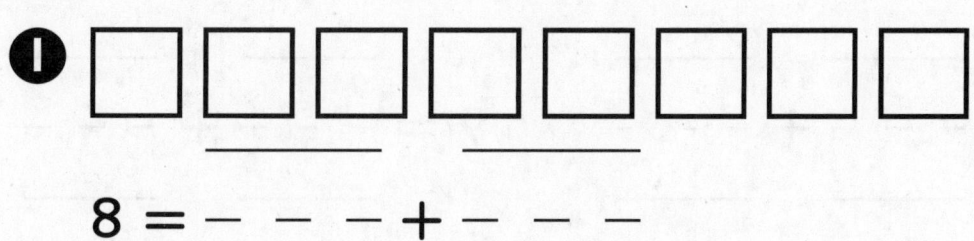

$8 = \_\ \_\ \_ + \_\ \_\ \_$

2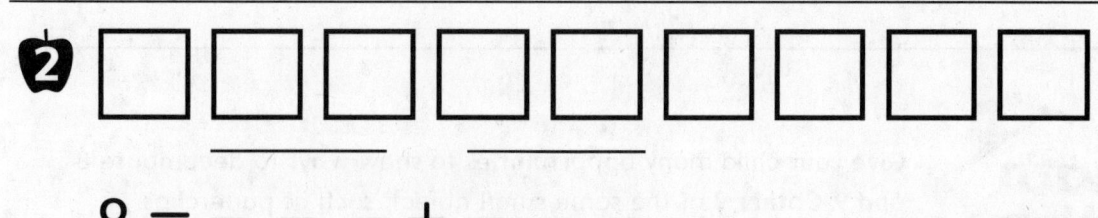

$9 = \_\ \_\ \_ + \_\ \_\ \_$

Review: You can decompose 8 and 9. The example shows two different ways to decompose 9.
Directions: 1. How can you decompose 8? Color each square either red or yellow. Write the equation to match. **2.** How can you decompose 9? Color each square either red or yellow. Write the equation to match.

$$8 = \underline{\quad} \overline{\underline{\quad}} + \underline{\quad} \overline{\underline{\quad}} \qquad 8 = \underline{\quad} \overline{\underline{\quad}} + \underline{\quad} \overline{\underline{\quad}}$$

$$9 = \underline{\quad} \overline{\underline{\quad}} + \underline{\quad} \overline{\underline{\quad}} \qquad 9 = \underline{\quad} \overline{\underline{\quad}} + \underline{\quad} \overline{\underline{\quad}}$$

Math @ Home Activity

Give your child many opportunities to show ways to decompose 8 and 9. Gather 9 of the same small object, such as paperclips. Have your child practice breaking apart groups of 8 or 9 objects.

Directions: 3. Draw counters to show 8. How can you decompose 8? Finish the equations to show two ways to decompose 8. **4.** Draw counters to show 9. How can you decompose 9? Finish the equations to show two ways to decompose 9.

Additional Practice

Name _____

Review

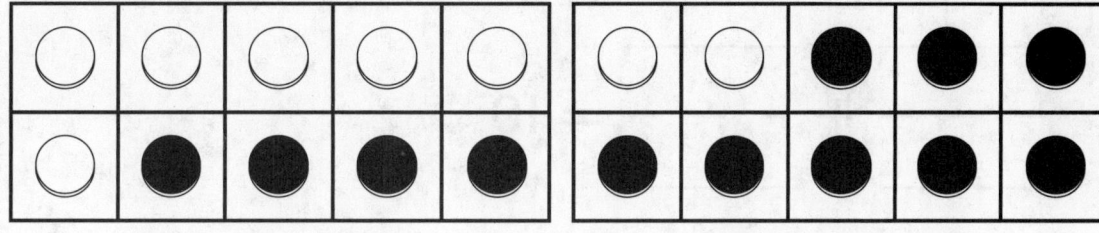

$$6 + 4 = 10 \qquad 2 + 8 = 10$$

1

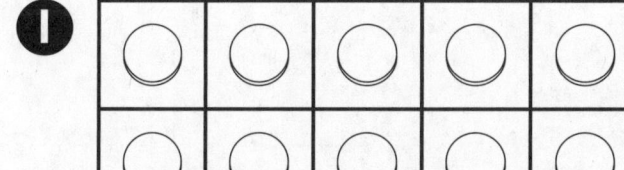

$$\underline{\quad\quad} + \underline{\quad\quad} = 10$$

 2

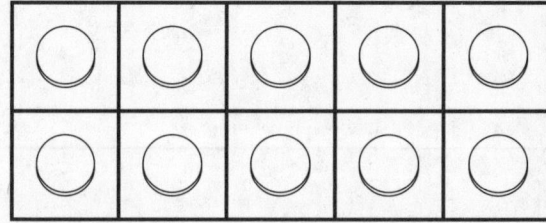

$$\underline{\quad\quad} + \underline{\quad\quad} = 10$$

Review: You can show ways to make 10. The example shows two different ways to make 10.

Directions: 1–2. How can you make 10? Use two colors to show a way to make 10. Write the equation to match.

3

_____ _____

‒ ‒ ‒ + ‒ ‒ ‒ = 10

_____ _____

4

_____ _____

‒ ‒ ‒ + ‒ ‒ ‒ = 10

_____ _____

**Math @ Home Activity**

Have your child create his or her own ten-frame on a sheet of paper. Using checker game pieces or similar objects in two colors, have your child practice making 10. Have him or her record an addition equation, such as 2 + 8 = 10, for each way to make 10.

Directions: 3. Draw 10 squares. Color to show one way to make 10. Write the equation to match. **4.** Draw 10 squares. Color to show a way to make 10 that is different from the way you colored the squares in Exercise 3. Write the equation to match.

Lesson 8-8

Additional Practice

Name _____

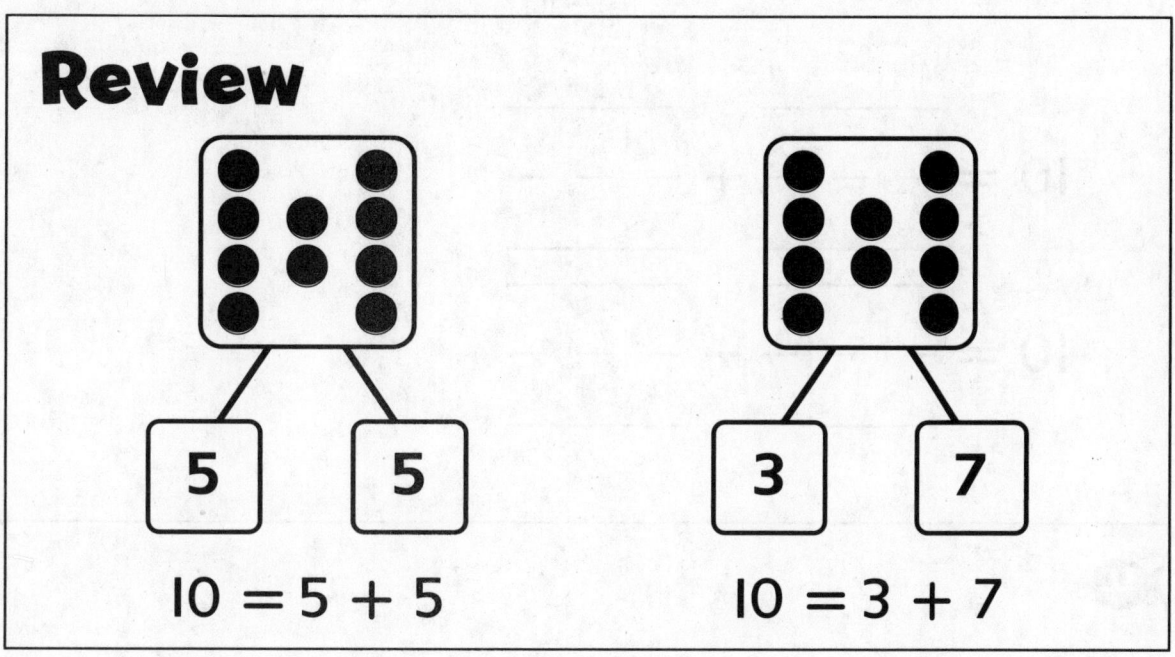

①

$$10 = \underline{\ \ } \underline{\ \ } \underline{\ \ } + \underline{\ \ } \underline{\ \ } \underline{\ \ }$$

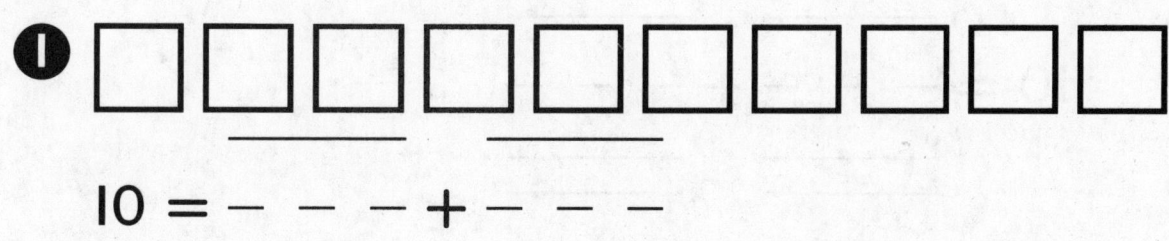

②

$$10 = \underline{\ \ } \underline{\ \ } \underline{\ \ } + \underline{\ \ } \underline{\ \ } \underline{\ \ }$$

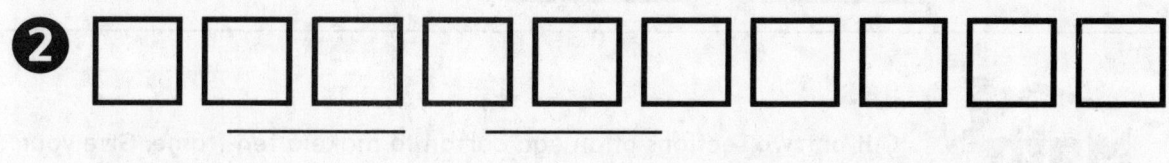

Review: You can decompose 10. The example shows two different ways to decompose 10.

Directions: 1. How can you decompose 10? Color each square either yellow or red. Write the equation to match. **2.** How can you decompose 10 in a different way? Color each square either yellow or red. Write the equation to match.

3

$$10 = \underline{} \underline{} \underline{} + \underline{} \underline{} \underline{}$$

$$10 = \underline{} \underline{} \underline{} + \underline{} \underline{} \underline{}$$

 4

$$10 = \underline{} \underline{} \underline{} + \underline{} \underline{} \underline{}$$

$$10 = \underline{} \underline{} \underline{} + \underline{} \underline{} \underline{}$$

Math @ Home Activity

Cut off two sections of an egg carton to make a ten-frame. Give your child green grapes and red grapes or similar objects. Have him or her practice breaking apart 10 using the different colored objects.

Directions: 3. Draw counters to show 10. How can you decompose 10? Finish the equations to show two ways to decompose 10. **4.** Draw counters to show 10. How can you decompose 10? Finish the equations to show two ways to decompose 10 that are different than the ways you showed in Exercise 3.

Student Practice Book

Additional Practice

Name _____

Review

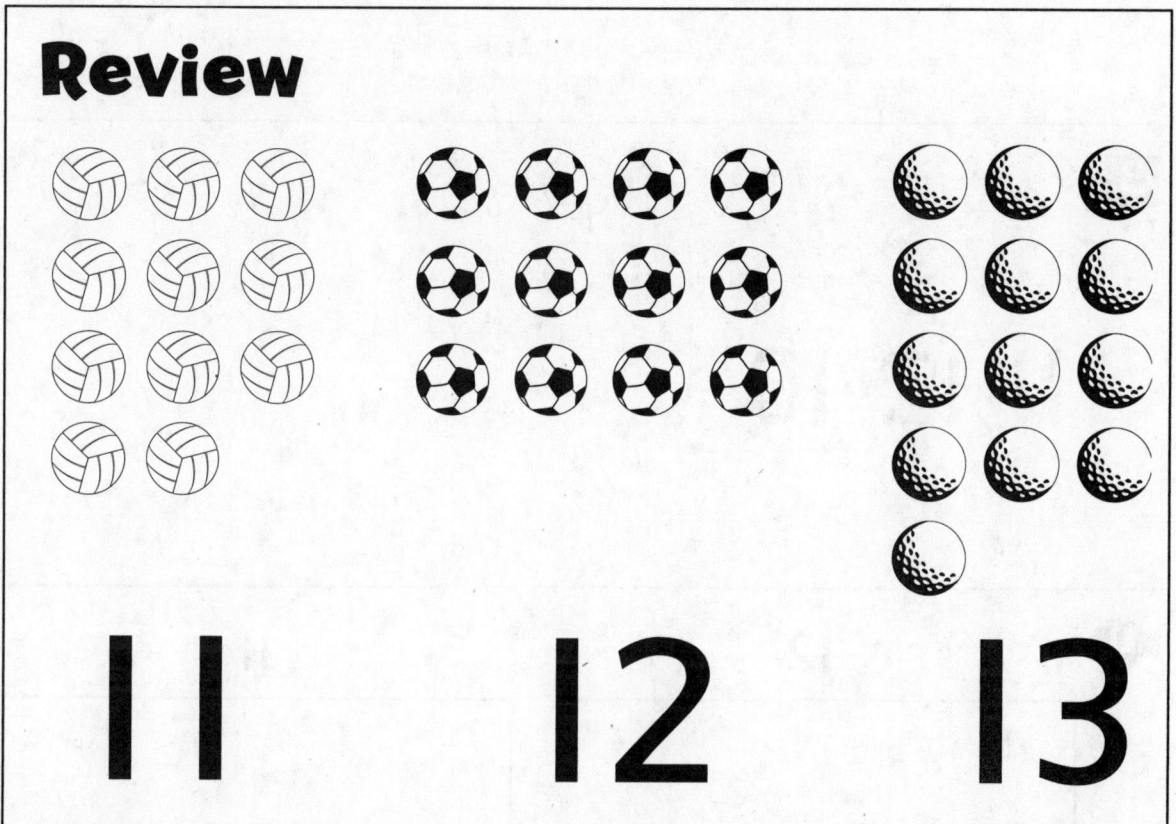

11   12   13

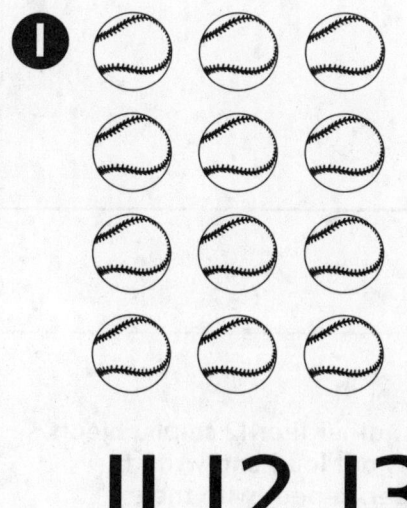

11 12 13

Review: You can count the number of objects and show how many.

Directions: 1. How many baseballs? Count. Circle the number to show how many.

2

11 12 13

3

11 12 13

---

4   12   11

[ ]   [ ]

Math @Home Activity

While playing outside, have your child gather 11 or 12 small objects. Your child should then count the objects out loud and write the number with chalk or on a sheet of paper. Repeat with the other number.

Directions: 2. How many whistles? Count. Circle the number to show how many. **3.** How many tennis balls? Count. Circle the number to show how many. **4.** Draw footballs to show each number.

Additional Practice

Name

Review

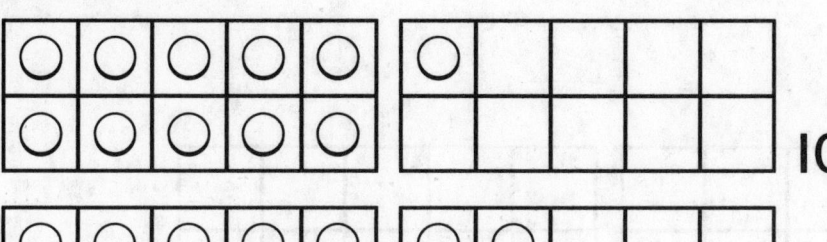

$$10 + 1 = 11$$

$$10 + 2 = 12$$

$$10 + 3 = 13$$

1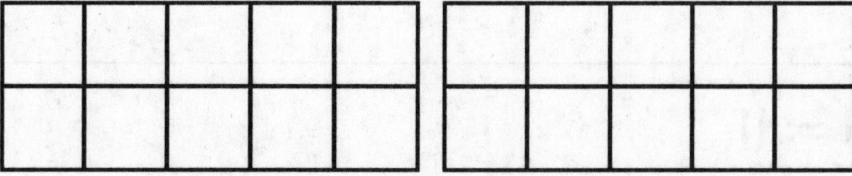

_____   _____

_ _ _ _ + _ _ _ _ = 12

_____   _____

Review: You can make 11, 12, and 13.

Directions: 1. How can you make 12? Draw counters to show a group of ten ones and some more ones. Write the equation to match.

2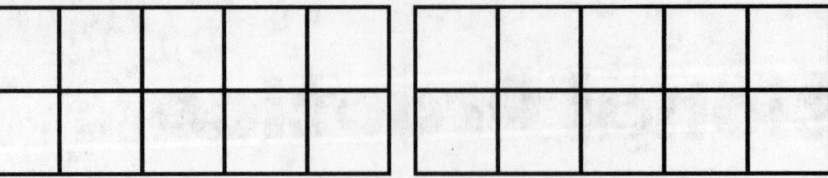

_____   _____

− − − − + − − − − = 11

_____   _____

3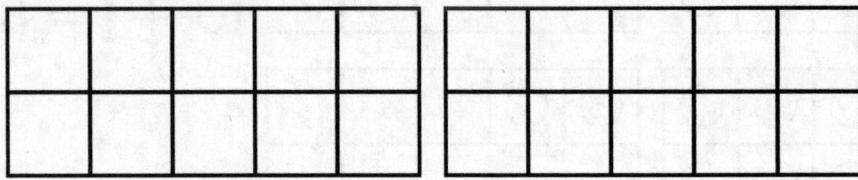

_____   _____

− − − − + − − − − = 13

_____   _____

4 $10 + 1 = 11$

Math @ Home Activity

Cut two sections off an egg carton to make a ten-frame. Give your child the egg carton ten-frame and the two-section portion of the carton, as well as some small objects such as pennies. Have him or her use the pennies and the egg carton pieces to show ways to make 11, 12, and 13. For example, your child can place 10 pennies in the ten-frame and 2 more pennies in the other egg carton piece to represent $10 + 2 = 12$.

Directions: 2. How can you make 11? Draw counters to show a group of ten ones and some more ones. Write the equation to match. **3.** How can you make 13? Draw counters to show a group of ten ones and some more ones. Write the equation to match. **4.** Draw a picture to show the equation $10 + 1 = 11$.

Student Practice Book

Additional Practice

Name _____

Review

$$13 = 10 + 3$$

1

$$11 = \underline{} \underline{} + \underline{} \underline{}$$

2

$$13 = \underline{} \underline{} + \underline{} \underline{}$$

Review: You can break apart 13 into ten ones and some more ones. 13 equals 10 plus 3.

Directions: 1. How can you decompose 11? Circle groups to decompose 11 into ten ones and some more ones. Write the equation to match. **2.** How can you decompose 13? Circle groups to decompose 13 into ten ones and some more ones. Write the equation to match.

3

$$12 = \underline{} + \underline{}$$

4 $13 = 10 + 3$

Math @ Home Activity

Draw a number bond on a sheet of paper. Have your child count and place 11 small objects in the top square of the number bond. Then have him or her break apart the number by moving 10 objects to the first small square and the rest of the objects to the other small square. Repeat with the numbers 12 and 13.

Directions: 3. How can you decompose 12? Circle groups to decompose 12 into ten ones and some more ones. Write the equation to match. **4.** Draw a picture to show the equation $13 = 10 + 3$.

Lesson 9-4

Additional Practice

Name _____

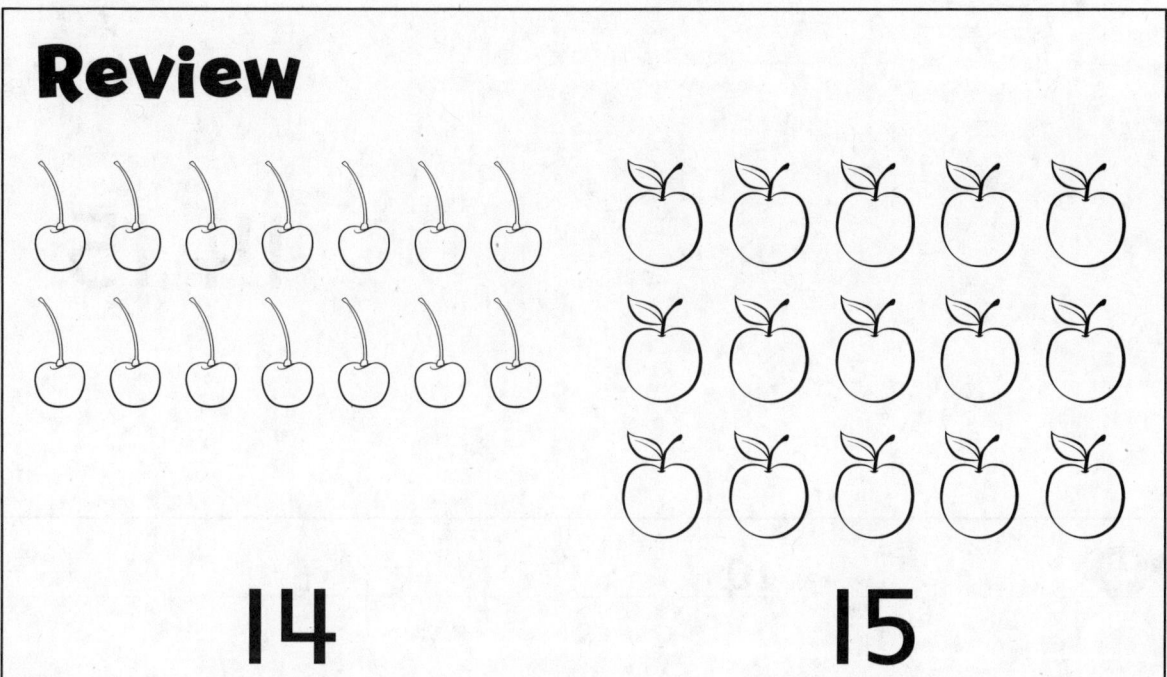

Review

14

15

1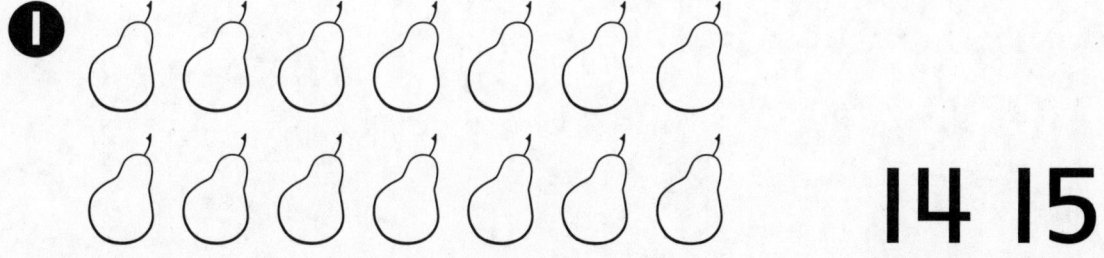

14 15

2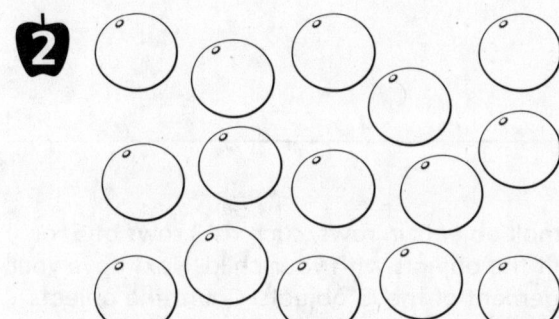

14 15

Review: You can count the number of objects and show how many.

Directions: **1.** How many pears? Count. Circle the number to show how many. **2.** How many oranges? Count. Circle the number to show how many.

Student Practice Book

3

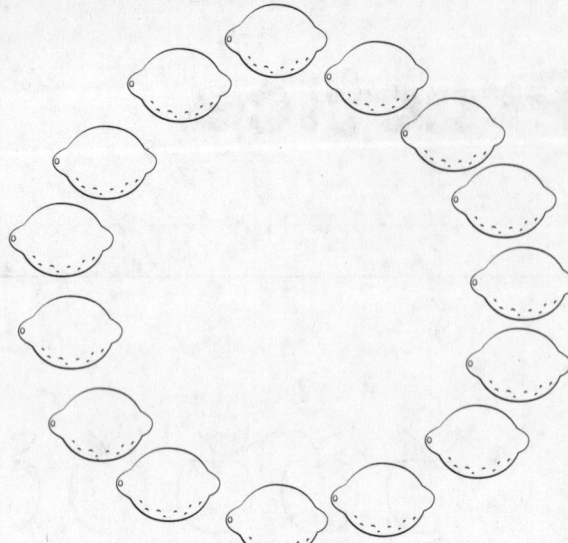

14 15

---

4   **14**

Math @ Home Activity

Have your child arrange 15 small objects in rows, such as 3 rows of 5, or 2 rows 6 and 1 row of 3. Count the objects with your child. Next have your child make a different arrangement of the 15 objects. Count the objects together. Discuss how, even though they have been arranged differently, there are still 15 objects. Repeat with the number 14.

Directions: 3. How many lemons? Count. Circle the number to show how many.
4. Draw bananas to show the number.

Additional Practice

Name _____

Review

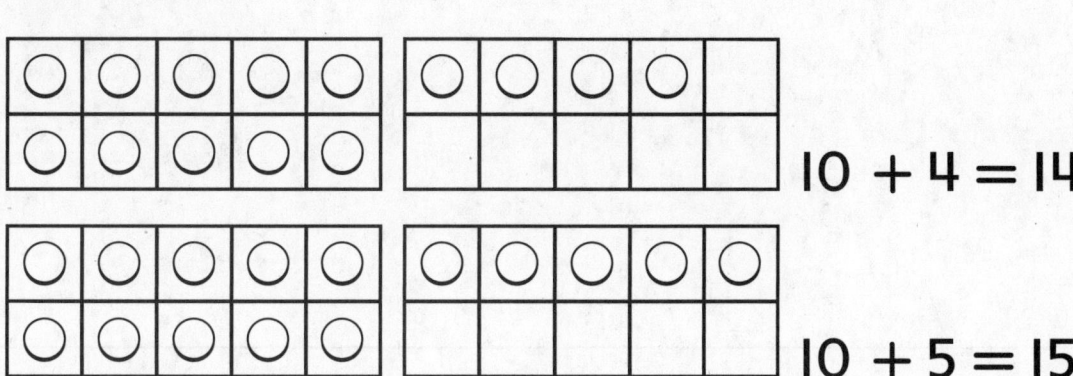

$$10 + 4 = 14$$

$$10 + 5 = 15$$

1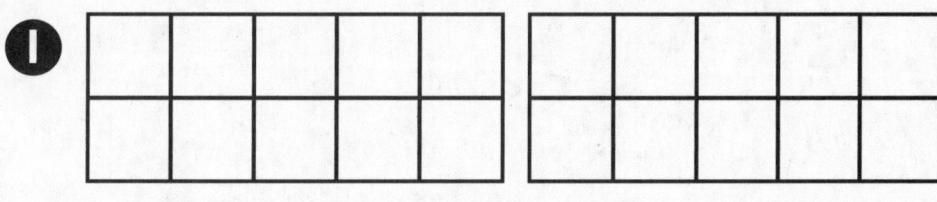

_____   _____

— — — — — + — — — — — = 14

_____   _____

2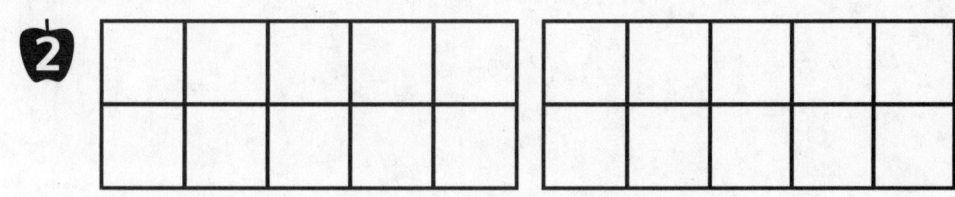

_____   _____

— — — — — + — — — — — = 15

_____   _____

Review: You can make 14 and 15.

Directions: 1. How can you make 14? Draw counters to show a group of ten ones and some more ones. Write the equation to match. **2.** How can you make 15? Draw counters to show a group of ten ones and some more ones. Write the equation to match.

3 $10 + 4 = 14$

---

4 $10 + 5 = 15$

Math @ Home Activity

Cut out 15 small triangles from a sheet of paper. Have your child show how to use 10 and some more triangles to make 15 and then write the corresponding equation ($10 + 5 = 15$). Repeat with 14 of the small triangles.

Directions: 3. Draw a picture to show the equation $10 + 4 = 14$. **4.** Draw a picture to show the equation $10 + 5 = 15$.

Additional Practice

Name _____

Review

$14 = 10 + 4$

$15 = 10 + 5$

---

❶

$$14 = \underline{} + \underline{}$$

---

❷

$$15 = \underline{} + \underline{}$$

Review: You can break apart 14 and 15 into ten ones and some more ones.

Directions: 1. How can you decompose 14? Circle groups to decompose 14 into ten ones and some more ones. Write the equation to match. **2.** How can you decompose 15? Circle groups to decompose 15 into ten ones and some more ones. Write the equation to match.

3

$$14 = \underline{\quad\quad} + \underline{\quad\quad}$$

4

$$15 = \underline{\quad\quad} + \underline{\quad\quad}$$

Copyright © McGraw-Hill Education

Math @ Home Activity

Write the number 14 on a sheet of paper. Have your child say, trace, and write the number. Then have him or her break apart the number. For example, have your child use 14 paper clips or other small objects to show how to use 10 and some more to break apart (14 = 10 + 4) the number. Repeat with the number 15.

Directions: 3. How can you decompose 14? Circle groups to decompose 14 into ten ones and some more ones. Write the equation to match. **4.** How can you decompose 15? Circle groups to decompose 15 into ten ones and some more ones. Write the equation to match.

Lesson 10-1

Additional Practice

Name _____

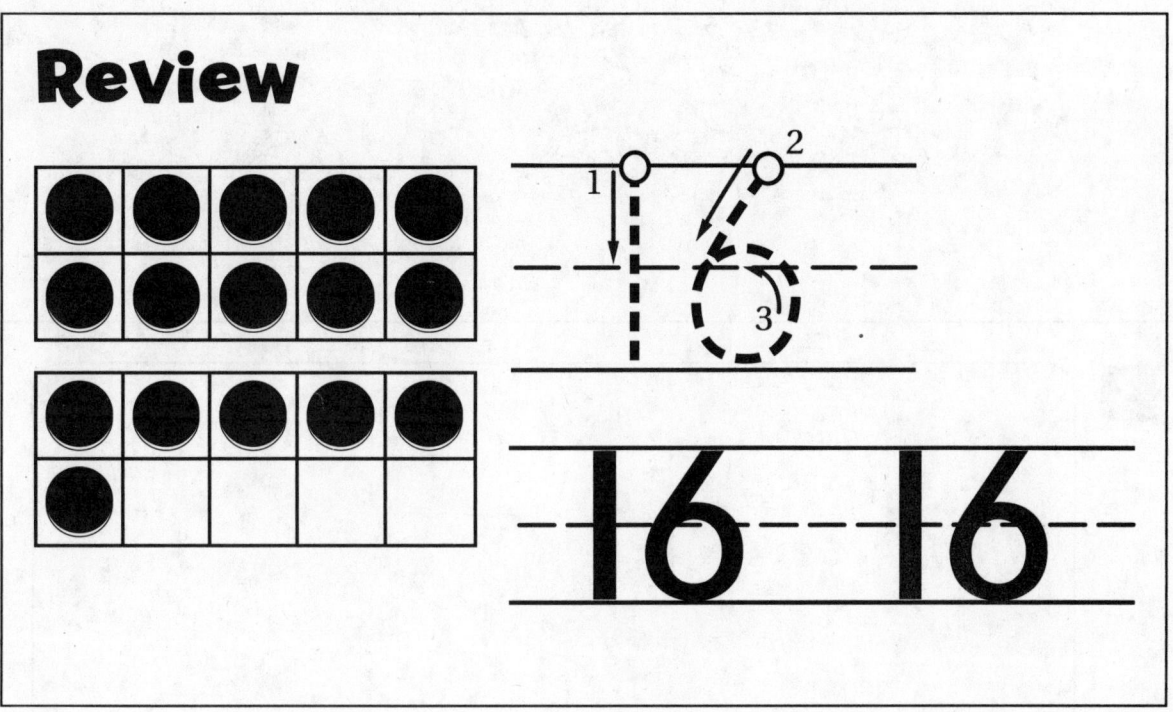

1

_____

- - - - - - -

_____

Review: You can use counters to make 16. You can trace and write 16 with a pencil.

Direction: 1. How many stuffed animals? Count. Write a number to show how many.

2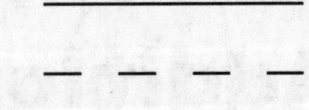

_ _ _ _ _ _

3

Math @ Home Activity

Encourage your child to practice counting, reading, and writing 16 and 17. Ask your child to write the numbers 16 and 17 at the top of a sheet of paper. Then have him or her draw a picture to represent each number.

Directions: 2. How many stuffed animals? Count. Write a number to show how many.
3. How can you show 17? Draw counters to show 17.

Lesson 10-2

Additional Practice

Name _____

Review

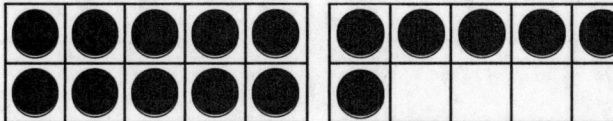

$10 + 6 = 16$

$10 + 7 = 17$

---

❶

⬜ ⬜

_____   _____   _____

___ ___ ___ **+** ___ ___ ___ **=** ___ ___ ___

_____   _____   _____

---

❷

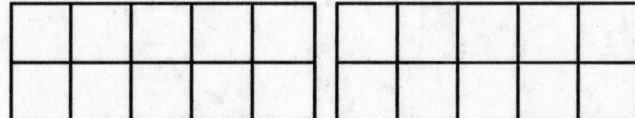

_____   _____   _____

___ ___ ___ **+** ___ ___ ___ **=** ___ ___ ___

_____   _____   _____

Review: You can make 16 and 17.

Directions: 1. How can you make 16? Draw counters to show a group of ten ones and some more ones. Write the equation to match. **2.** How can you make 17? Draw counters to show a group of ten ones and some more ones. Write the equation to match.

Student Practice Book

III

3

$$\underline{10} \quad + \quad \text{-----} \quad = \quad \underline{17}$$

4

$$\text{-----} \quad + \quad \text{-----} \quad = \quad \underline{16}$$

Math @ Home Activity

Have your child practice making 16 and 17. Cut out 17 small squares from a sheet of paper. Have your child use the small squares to show making 16 and 17.

Directions: 3. Nancy says there are 10 books and 6 more books. Rochelle says there are 10 books and 7 more books. Complete the equation to show how many more books there are. **4.** Use one color to show 10 ones. Then use another color to show 6 more ones. Write the equation to match.

Additional Practice

Name

Review

$16 = 10 + 6$

$17 = 10 + 7$

①

$-16- = ----- + -----$

Review: You can break apart 16 and 17.

Directions: 1. How can you decompose 16? Circle groups to decompose 16 into ten ones and some more ones. Write the equation to match.

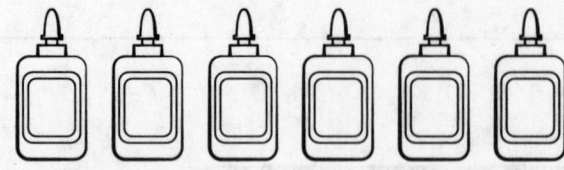

__ 16 __ = __ ___ __ + __ ___ __

3

__ 17 __ = __ ___ __ + __ ___ __

Copyright © McGraw-Hill Education

Math @ Home Activity

Have your child practice breaking apart 16 and 17. Cut out 17 small triangles from a sheet of paper. Have your child use the small triangles to show breaking apart 16 and 17.

Directions: 2. How can you decompose 16? Circle groups to decompose 16 into ten ones and some more ones. Write the equation to match. **3.** How can you decompose 17? Circle groups to decompose 17 into ten ones and some more ones. Write the equation to match.

Additional Practice

Name _____

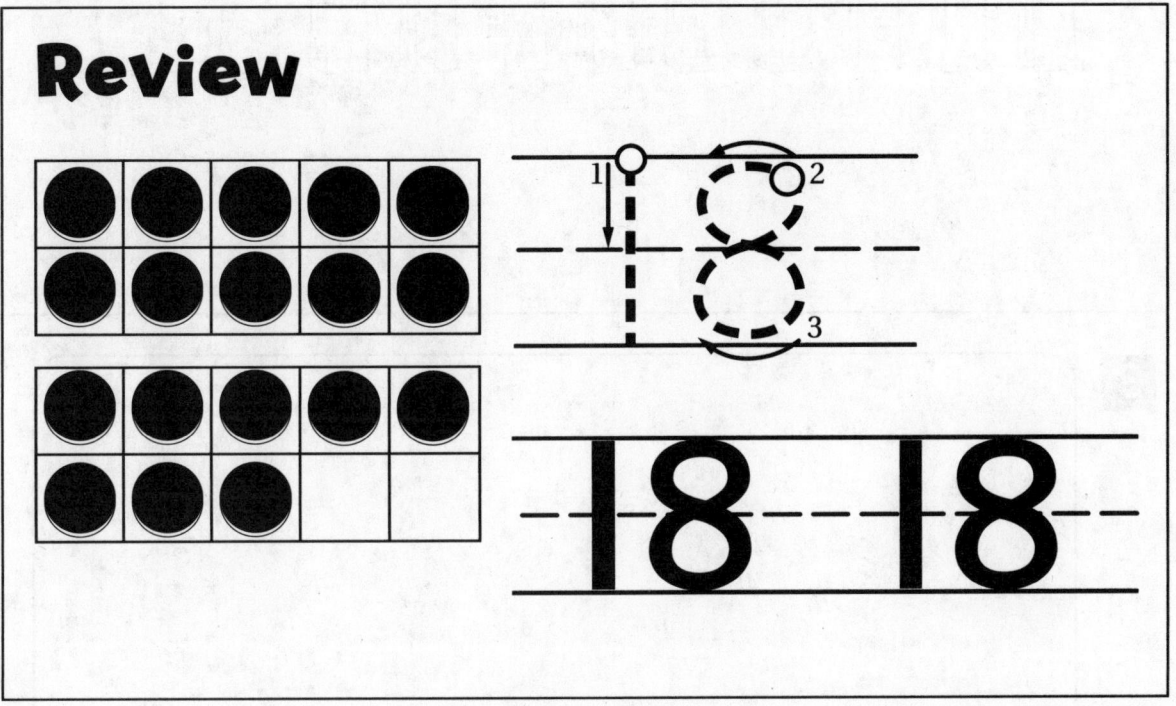

Review

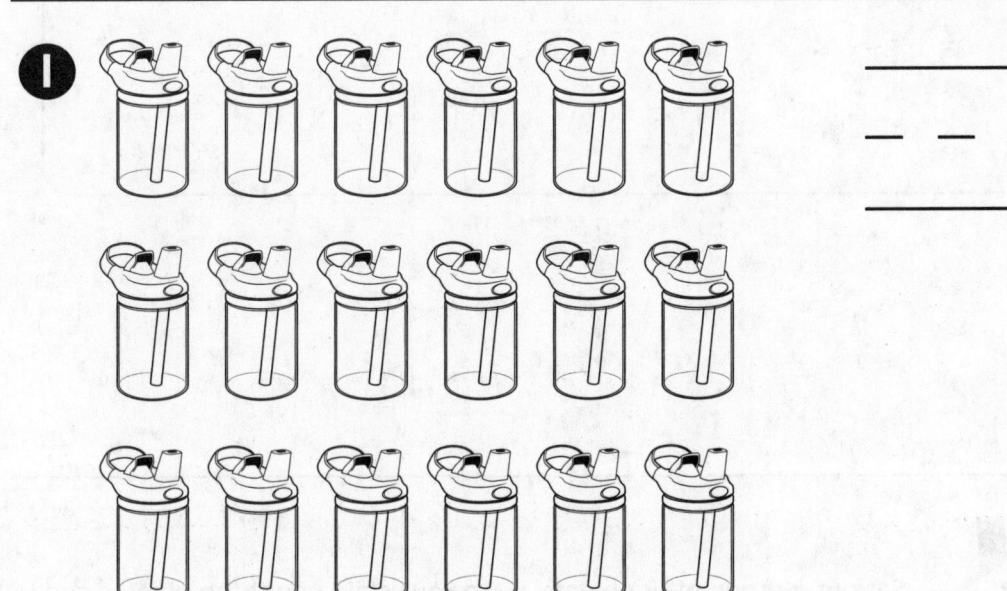

1.

_____

- - - - - - - - - - - -

_____

Review: You can use counters to make 18. You can trace and write 18.

Direction: 1. How many water bottles? Count. Write a number to show how many.

2

_____

- - - - - - -

_____

3

Math @ Home Activity

Set out a group of 18 objects. Have your child count the objects and write how many there are. Encourage him or her to practice writing the numbers. Repeat the activity with 19 objects.

Directions: 2. How many bags? Count. Write a number to show how many.
3. How can you show 18? Draw counters to show 18.

Additional Practice

Name _____

Review

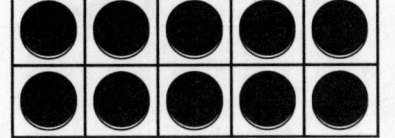 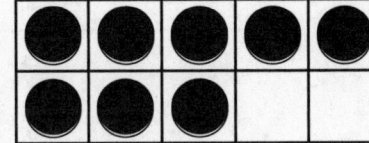 $10 + 8 = 18$

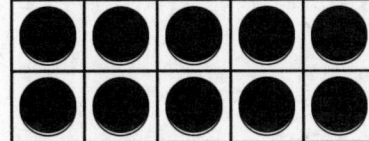  $10 + 9 = 19$

---

❶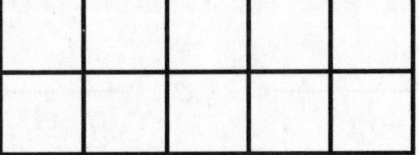

_____ $+$ _____ $=$ _____

---

❷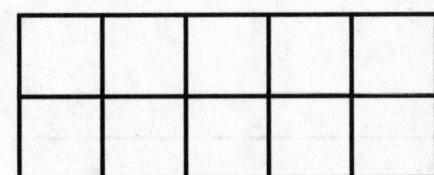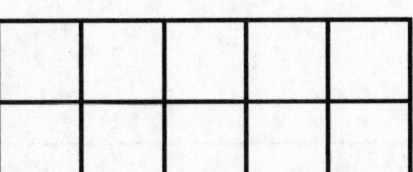

_____ $+$ _____ $=$ _____

Review: You can make 18 and 19.

Directions: 1. How can you make 18? Draw counters to show a group of ten ones and some more ones. Write the equation to match. **2.** How can you make 19? Draw counters to show a group of ten ones and some more ones. Write the equation to match.

3

$$\text{10} + \underline{} = \text{18}$$

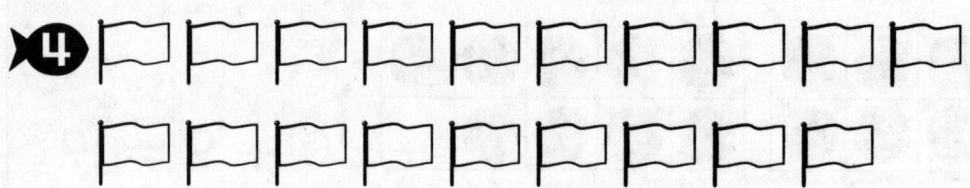

$$\underline{} + \underline{} = \text{19}$$

Math @ Home Activity

Have your child use small objects to show how to break apart 18. Then help him or her write an addition sentence for breaking apart 18. Repeat the activity with breaking apart 19.

Directions: 3. Landon says there are 10 gifts and 8 more gifts. Ashton says there are 10 gifts and 9 more gifts. Complete the equation to show how many more gifts there are.
4. Use one color to show 10 ones. Then use another color to show 9 more ones. Write the equation to match.

Lesson 10-6

Additional Practice

Name ______________________

Review

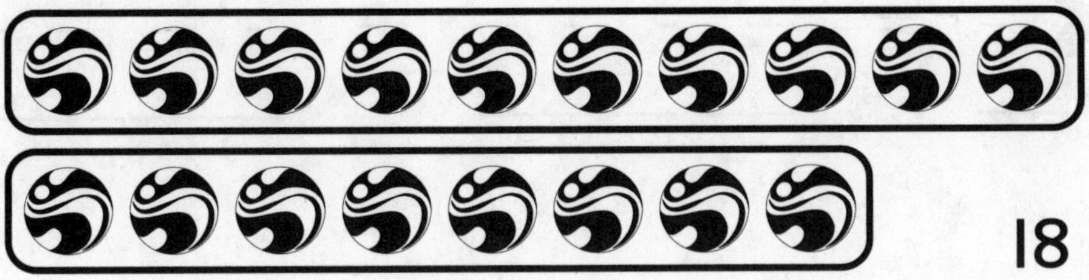

$18 = 10 + 8$

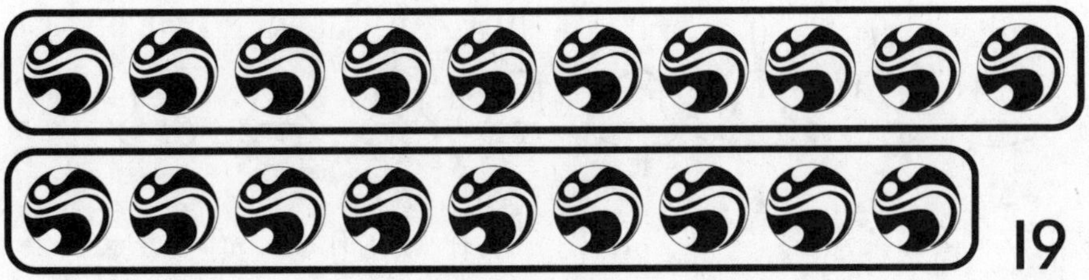

$19 = 10 + 9$

1

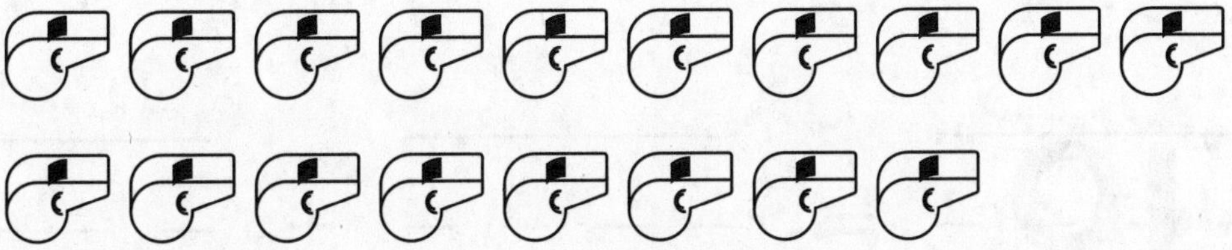

18 = ______ + ______

Review: You can break apart 18 and 19.

Directions: 1. How can you decompose 18? Circle groups to decompose 18 into ten ones and some more ones. Write the equation to match.

2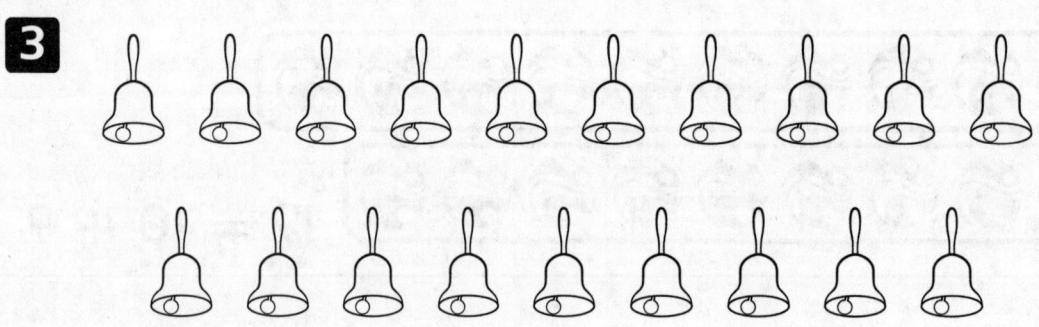

$$\overline{18} = \overline{} + \overline{}$$

3

$$\overline{19} = \overline{} + \overline{}$$

Math @ Home Activity

Have your child practice breaking apart 18 and 19. Cut out 19 small rectangles from a sheet of paper. Have your child use the small rectangles to show breaking apart 18 and 19.

Directions: 2. How can you decompose 18? Circle groups to decompose 18 into ten ones and some more ones. Write the equation to match. **3.** How can you decompose 19? Circle groups to decompose 19 into ten ones and some more ones. Write the equation to match.

Additional Practice

Name _____

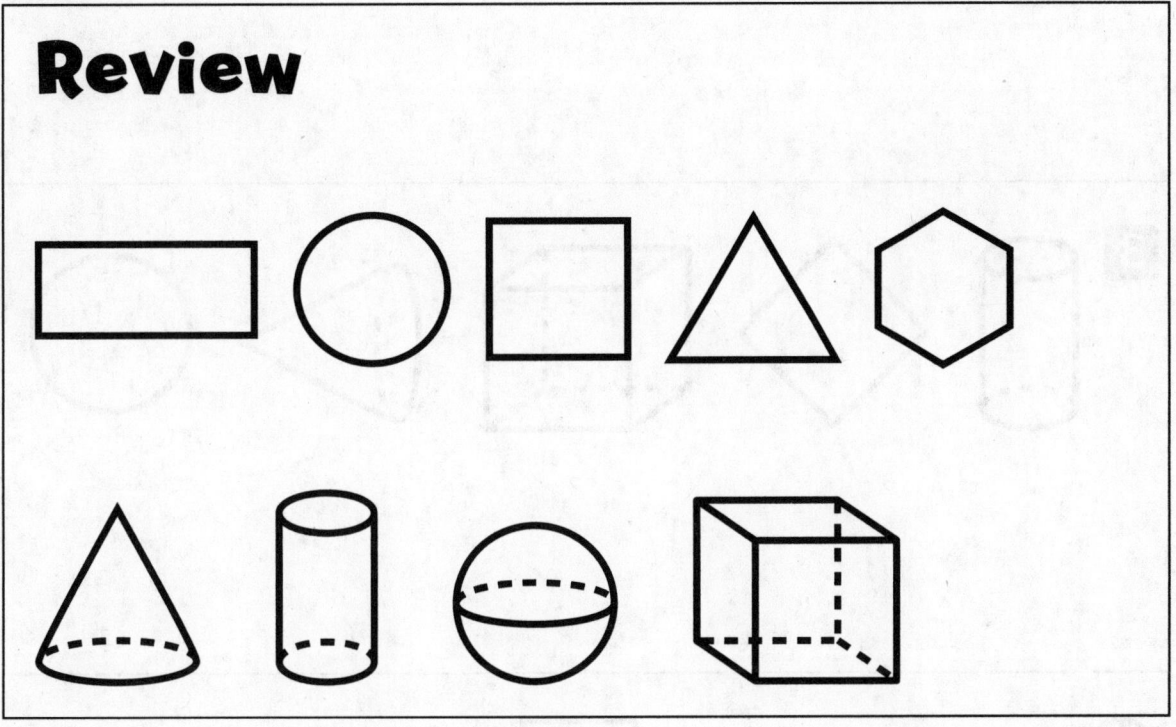

Review

---

①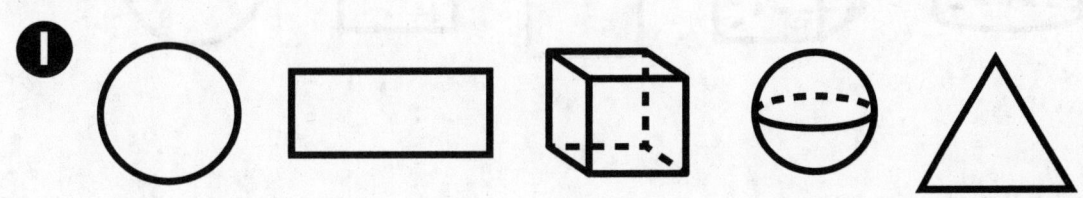

Review: You can identify shapes as 2-dimensional or 3-dimensional.
Directions: 1. Which shapes are 3-dimensional? Circle all the 3-dimensional shapes.

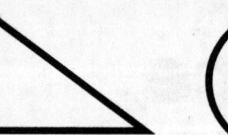  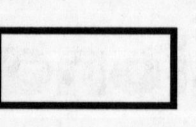 

   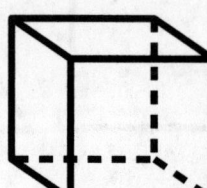  

Math @ Home Activity

While sitting in a room in your home, ask your child to point out objects that are 2-dimensional. Then have your child point out objects that are 3-dimensional. Discuss the differences between 2-dimensional and 3-dimensional shapes.

Directions: 2. Which shapes are 2-dimensional? Circle all the 2-dimensional shapes.
3. Which shapes are 3-dimensional? Circle all the 3-dimensional shapes.
4. Which shapes are 2-dimensional? Circle all the 2-dimensional shapes.

Student Practice Book

Additional Practice

Name _____

Review

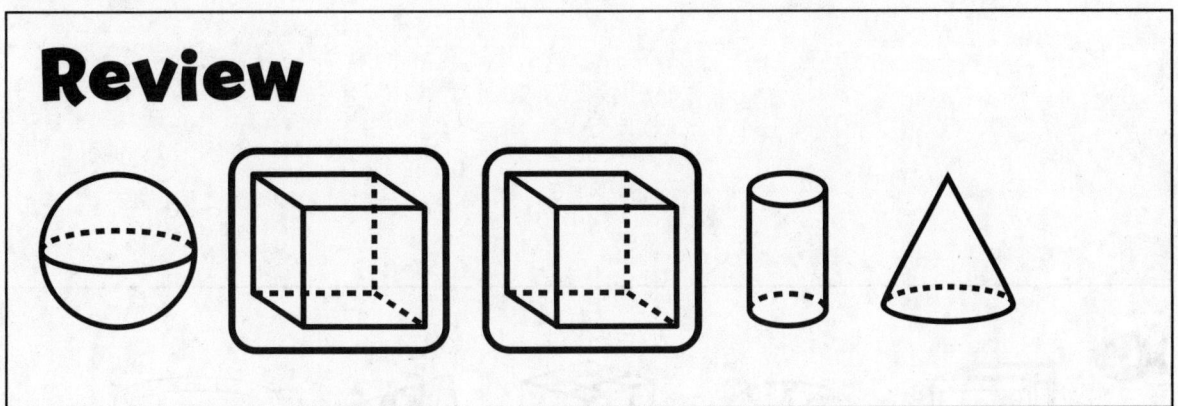

1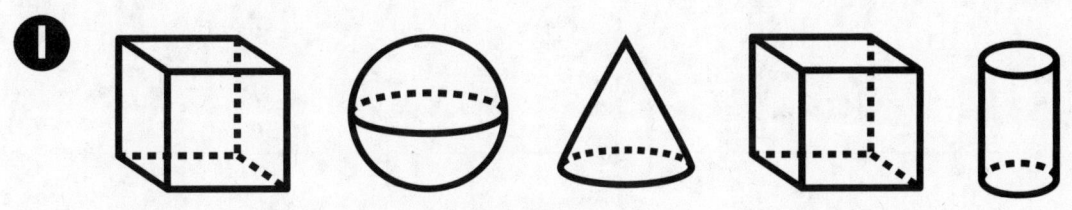

2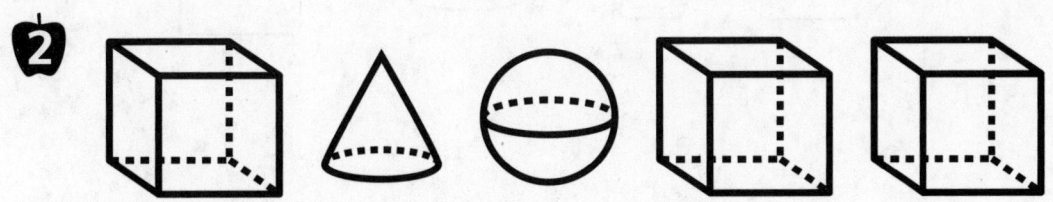

Review: You can identify cubes.

Directions: I–2. Which shapes are cubes? Circle all of the cubes.

3

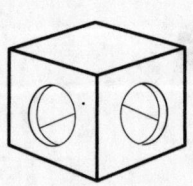   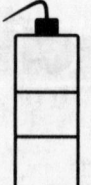 

4

5

 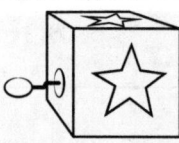 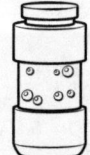 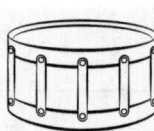 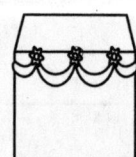

Math @ Home Activity

While putting away the toys in your home, ask your child to identify toys that are shaped like cubes. Your child may point out blocks, a jack-in-the-box, etc.

Directions: 3–5. Which objects are shaped like cubes? Circle all of the objects shaped like cubes.

Lesson 11-3

Additional Practice

Name _____

Review

  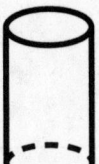 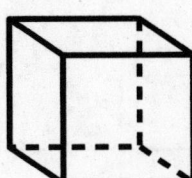 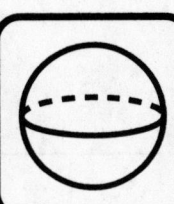

①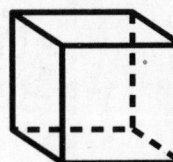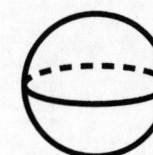

②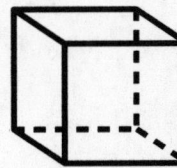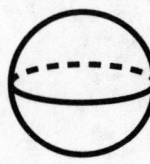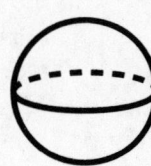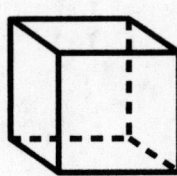

Review: You can identify spheres.

Directions: 1–2. Which shapes are spheres? Circle all of the spheres.

3

4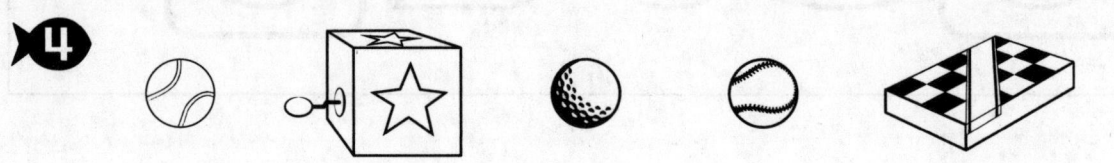

5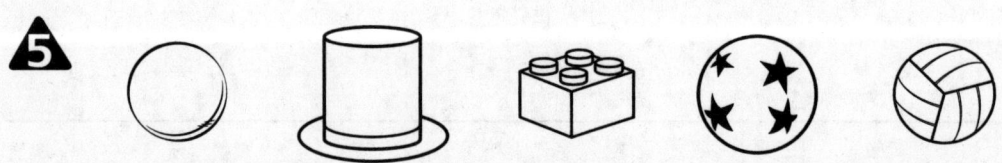

Math @ Home Activity

Give your child many opportunities to identify spheres. While walking through the park or neighborhood, ask your child to identify objects that are shaped like spheres. Point out other 3-dimensional shapes, such as cubes and cylinders, and ask your child to explain why they are not spheres.

Directions: 3–5. Which objects are shaped like spheres? Circle all of the objects shaped like spheres.

Lesson 11-4

Additional Practice

Name _____

Review

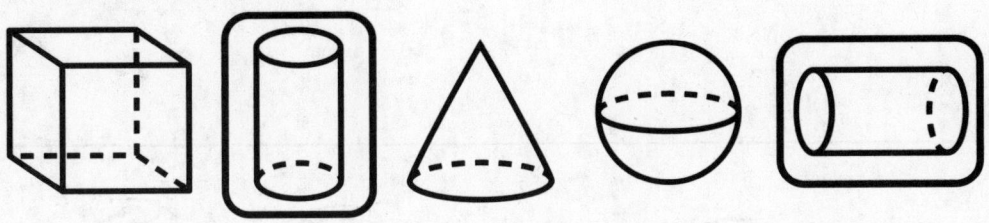

1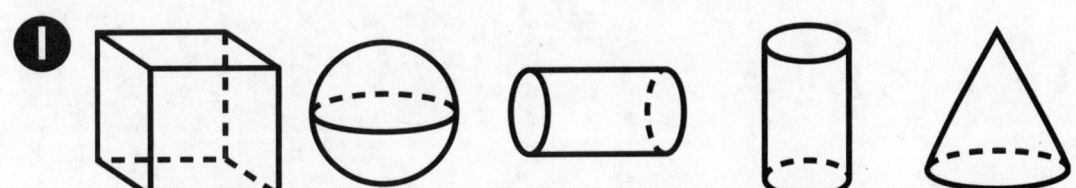

2

Review: You can identify cylinders.

Directions: 1–2. Which shapes are cylinders? Circle all of the cylinders.

Student Practice Book

127

3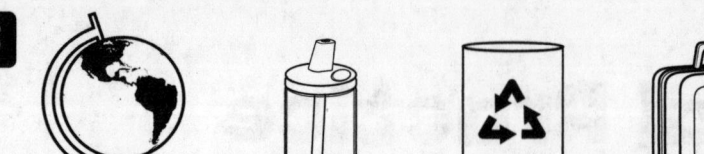

4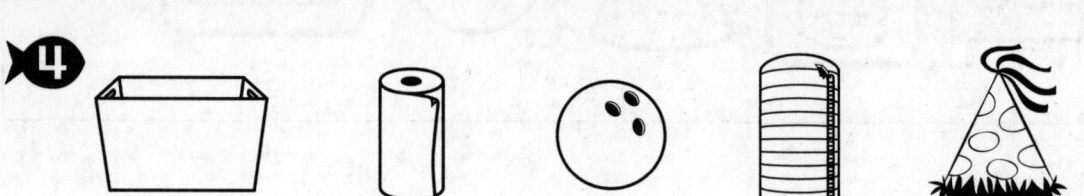

5

Math @ Home Activity

Have your child explain how to identify a cylinder. While walking around a grocery store, ask your child to identify objects that are shaped like a cylinder. Ask your child to explain why each identified object is a cylinder.

Directions: 3–5. Which objects are shaped like cylinders? Circle all of the objects shaped like cylinders.

Additional Practice

Name

Review

  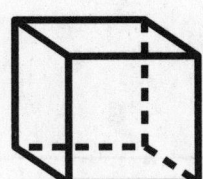 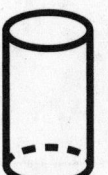 

1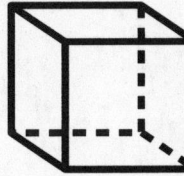

2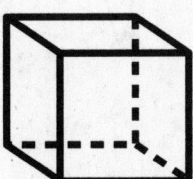

Review: You can identify cones.

Directions: 1–2. Which shapes are cones? Circle all of the cones.

3

 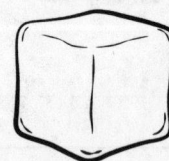   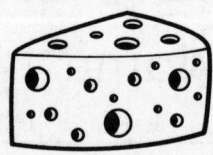

 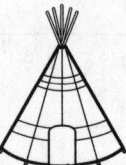    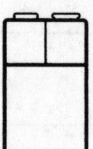

5

Math @ Home Activity

Cut a circle out of a piece of paper, and then cut a slit from the edge of the circle to the center. Give your child the paper circle, and ask your child to make a cone. Discuss how a flat circle should be added to the open space at the bottom to make it a true cone.

Directions: 3–5. Which objects are shaped like cones? Circle all of the objects shaped like cones.

Additional Practice

Name _____

Review

1

2

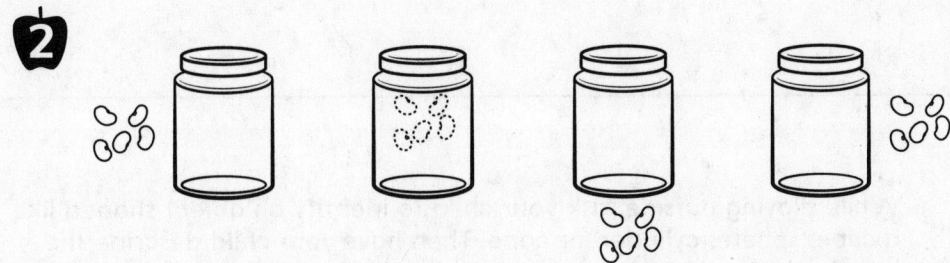

Review: You can identify the picture that shows the ice cream cone above the ant.

Directions: 1. How can you find an object that is *below*? Circle the picture that shows the toy box below the stuffed snake. **2.** How can you find an object that is *in front of*? Circle the picture that shows the jar in front of the beans.

3

4

Directions: 3. How can you find an object that is *next to*? Circle the picture that shows the ball next to the racket. **4.** How can you find an object that is *behind*? Circle the picture that shows the party hat behind the cake.

Additional Practice

Name _____

Review

1	2	3	4	5	6	7	8	9	10
11	12	13	14	15	16	17	18	19	20
21	22	23	24	25	26	27	28	29	30
31	32	33	34	35	36	37	38	39	40
41	42	43	44	45	46	47	48	49	50

❶

21	22	23	24				28	29	30

❷ 48, 47, 46   48, 46, 44   48, 49, 50

Review: You can count from 1 to 50 using a number chart. A pattern you can see in the third row is that all but the last number begin with 2 and the word *twenty*.

Directions: 1. Which numbers are missing? Count. Write the missing numbers. **2.** Start at 48. Circle the numbers you say when counting by 1s.

3

2_	2_	2_	24	25	26	27	28	29	30
31	32	33	34	35	36	37	38	39	40

4

31	32	33	34	35	3_	3_	3_	39	40
41	42	43	44	45	46	47	48	49	50

Math @ Home Activity

Write a random number between 1 and 47 on a sheet of paper. Then have your child write the next three numbers. Repeat the activity with different starting numbers.

Directions: 3–4. Which numbers are missing? Count. Write the missing numbers.

Additional Practice

Name _____

Review

1	2	3	4	5	6	7	8	9	10
11	12	13	14	15	16	17	18	19	20
21	22	23	24	25	26	27	28	29	30
31	32	33	34	35	36	37	38	39	40
41	42	43	44	45	46	47	48	49	50
51	52	53	54	55	56	57	58	59	60
61	62	63	64	65	66	67	68	69	70
(71)	72	73	74	75	76	77	78	79	80
81	82	83	84	85	86	87	88	89	90
91	92	93	94	95	96	97	98	99	100

1 | | 92 | | 94 | 95 | | 97 | | 99 | 100 |

Review: You can count from 1 to 100 using the chart. The number 71 is circled. You can use the chart to see that 72 comes after 71.

Directions: 1. Which numbers are missing? Count. Write the missing numbers.

2 53, 54, 55   53, 55, 57   53, 52, 51

3

81	8_	8_	8_	85	86	87	88	89	90
91	92	93	94	95	96	97	98	99	100

4

61	62	63	64	65	6_	6_	6_	69	70
71	72	73	74	75	76	77	78	79	80

Math @ Home Activity

Give your child opportunities to count to 100 by 1s. Start counting from a random number. When you stop counting, have your child continue counting on. Take turns choosing a starting number and counting on.

Directions: 2. Start at 53. Circle the numbers you say when counting by 1s. **3–4.** Which numbers are missing? Count. Write the missing numbers.

Additional Practice

Name _____

Review

1	2	3	4	5	6	7	8	9	10
11	12	13	14	15	16	17	18	19	20
21	22	23	24	25	26	27	28	29	30
31	32	33	34	35	36	37	38	39	40
41	42	43	44	45	46	47	48	49	50
51	52	53	54	55	56	57	58	59	60
61	62	63	64	65	66	67	68	69	70
71	72	73	74	75	76	77	78	79	80
81	82	83	84	85	86	87	88	89	90
91	92	93	94	95	96	97	98	99	100

1

1	2	3	4	5	6	7	8	9	
11	12	13	14	15	16	17	18	19	20
21	22	23	24	25	26	27	28	29	
31	32	33	34	35	36	37	38	39	40
41	42	43	44	45	46	47	48	49	

Review: You can touch and count by 10s to 100. A pattern you can see in the last column is that all the numbers end with 0.

Directions: 1. Which numbers are missing? Count. Write the missing numbers.

2

▢▢▢▢▢▢▢▢▢▢
▢▢▢▢▢▢▢▢▢▢

10   20   30

3

▢▢▢▢▢▢▢▢▢▢
▢▢▢▢▢▢▢▢▢▢
▢▢▢▢▢▢▢▢▢▢
▢▢▢▢▢▢▢▢▢▢
▢▢▢▢▢▢▢▢▢▢
▢▢▢▢▢▢▢▢▢▢
▢▢▢▢▢▢▢▢▢▢
▢▢▢▢▢▢▢▢▢▢

80   90   100

Math @ Home Activity

Write the numbers 10, 20, 30, 40, 50, 60, 70, 80, 90, and 100 on small pieces of paper. Mix up the pieces of paper and have your child arrange them in order while counting to 100 by 10s.

Directions: 2–3. How many? Count by 10s. Circle the number that shows how many.

Additional Practice

Name _____

Review

1	2	3	4	5	6	7	8	9	10
11	12	13	14	15	16	17	18	19	20
21	22	23	24	25	26	27	(28)	29	30
31	32	33	34	35	36	37	38	39	40
41	42	43	44	45	46	47	48	49	50
51	52	53	54	55	56	57	58	59	60
61	62	63	64	65	66	67	68	69	70
71	72	73	74	75	76	77	78	79	80
81	82	83	84	85	86	87	88	89	90
91	92	93	94	95	96	97	98	99	100

1. **42 43 44 _ _ _ _**

Review: You can touch and count by starting at any number. If you start at 28 and count to 32, you say the numbers 28, 29, 30, 31, 32.

Directions: 1. What number comes next? Count. Write the missing number.

2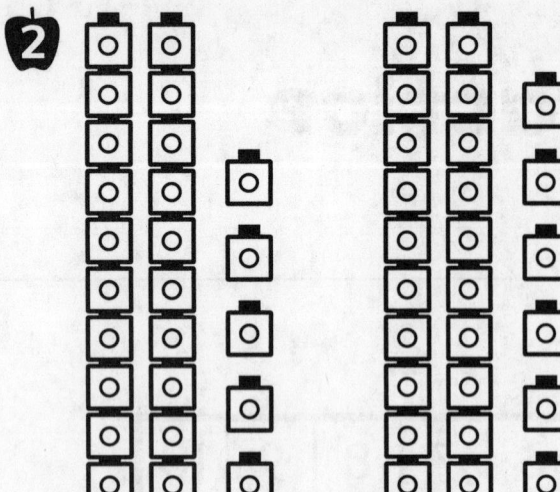

_____

- - - - - - -

_____

3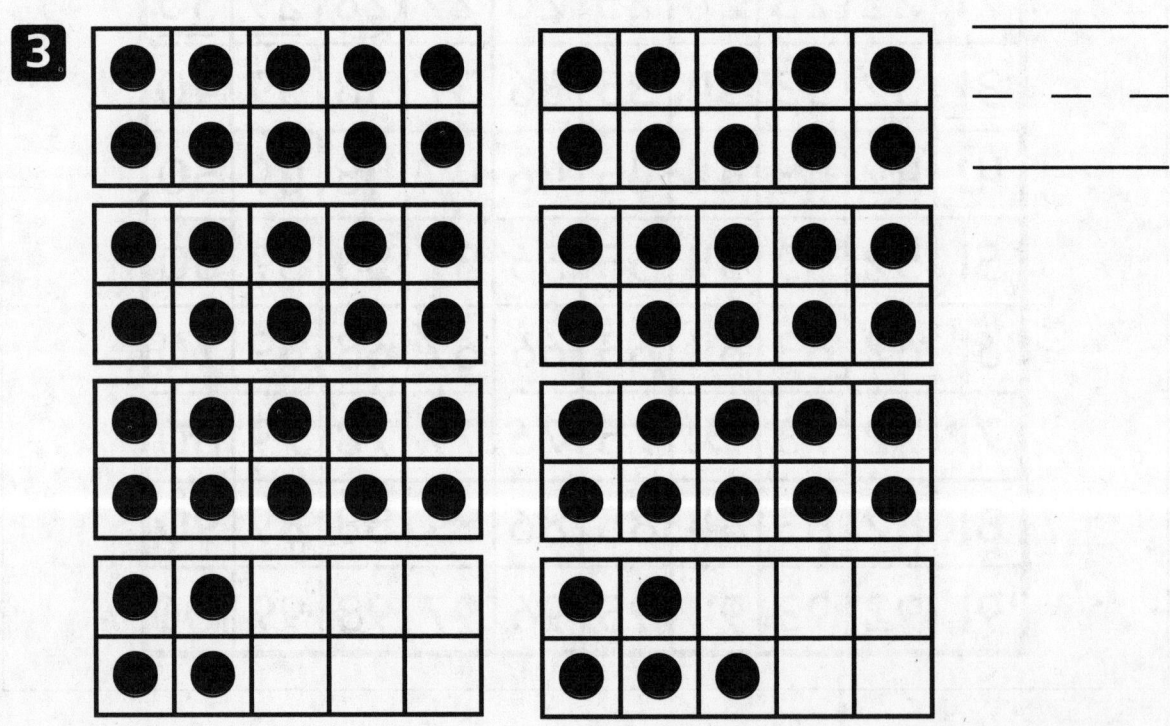

_____

- - -

_____

Math @ Home Activity

Write a number from 1 to 95 for your child. Have your child write the next 5 numbers. Check your child's work by touching and saying each number aloud. Then switch roles and have your child check your work.

Directions: 2. How many? Count each group of cubes. Write the number that comes next. **3.** How many? Count each group of counters. Then write the number that comes next.

Additional Practice

Name

Review

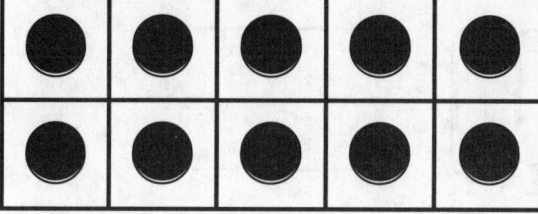 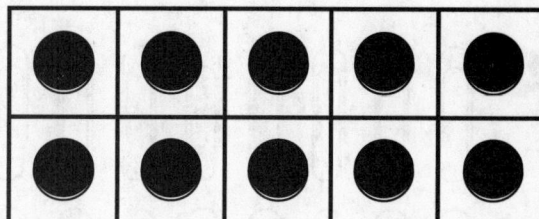

20

---

1  _____

- - - - -

_____

---

2 _____

- - - - -

_____

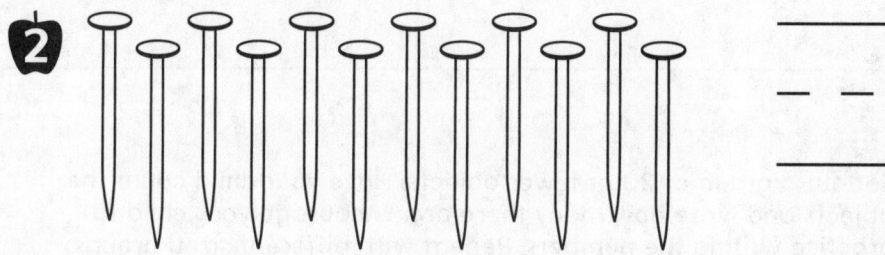

Review: You can use counters to make 20.

Directions: 1. How many? Count. Write the number to show how many bottles of nail polish. **2.** How many? Count. Write the number to show how many nails.

Student Practice Book

3

_____
- - - - - - -
_____

_____

4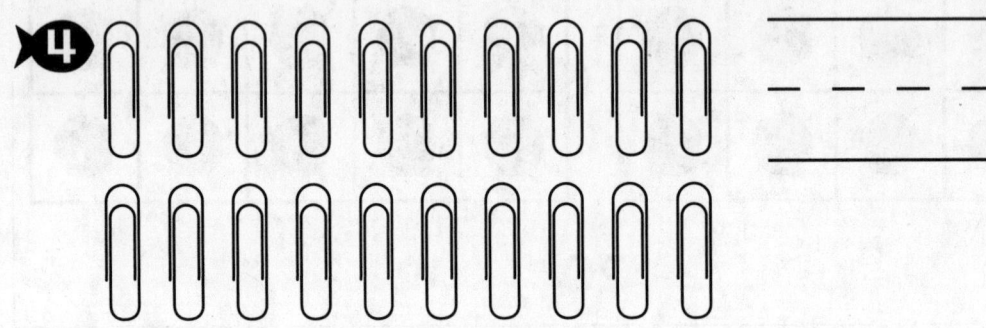

_____
- - - - - - -
_____

Math @ Home Activity

Set out a group of 20 or fewer objects. Have your child count the objects and write how many there are. Encourage your child to practice writing the numbers. Repeat with different sized groups.

Directions: 3. How many? Count. Write the number to show how many diamonds. **4.** How many? Count. Write the number to show how many paper clips.

Additional Practice

Name _____

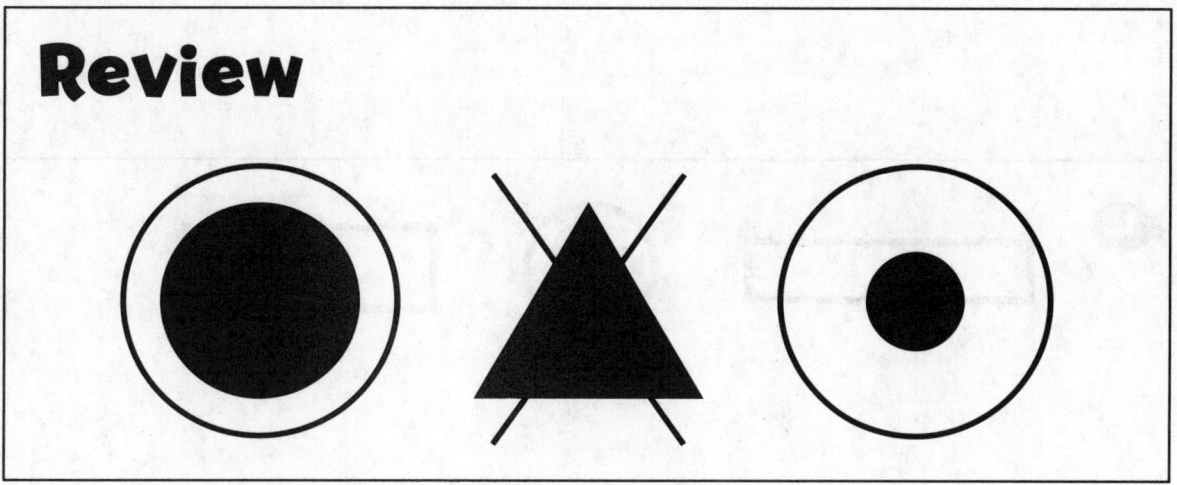

Review

1

2

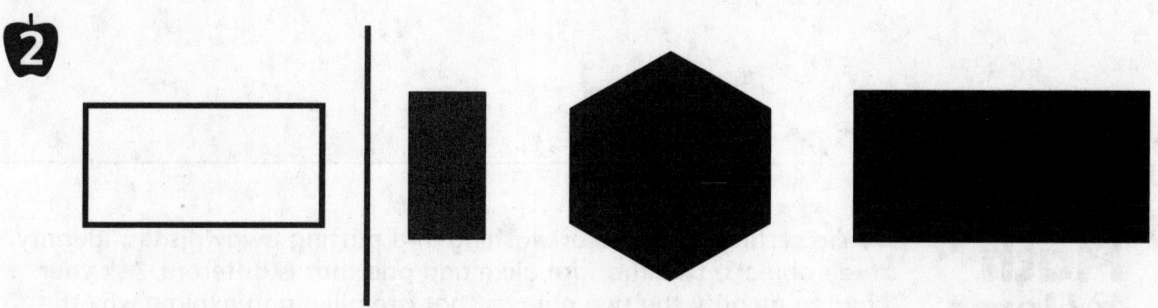

Review: You can compare shapes by counting the sides and vertices. The shapes that are circled are alike. The shape with an X on it is different.

Directions: 1. Which shapes are alike? Circle the shapes that are alike. **2.** How can you compare the shapes? Circle the shape that is different from the first shape.

Student Practice Book

3

4

Copyright © McGraw-Hill Education

Math @ Home Activity

While setting the table or washing and putting away dishes, identify three objects, two that are alike and one that is different. Ask your child to identify the two objects that are alike and explain why the third object is not like the others.

Directions: 3. Which shape is different? Draw an X on the shape that is different.
4. Which shapes are alike? Circle the shapes that are alike.

Additional Practice

Name _____

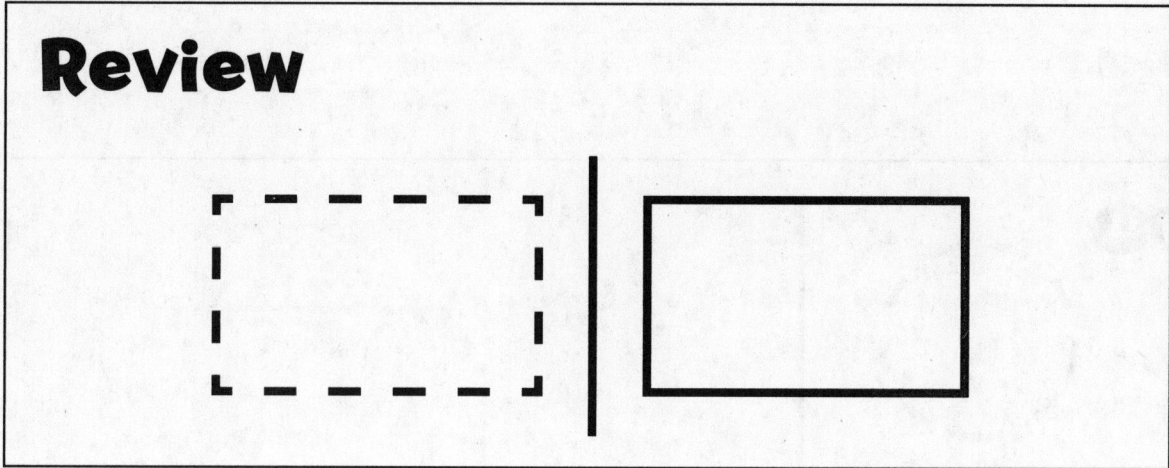

Review

---

①

---

②

Review: You can trace and draw shapes.

Directions: 1. How can you draw a triangle? Trace the triangle. Then draw a triangle.
2. How can you draw a hexagon? Trace the hexagon. Then draw a hexagon.

3

4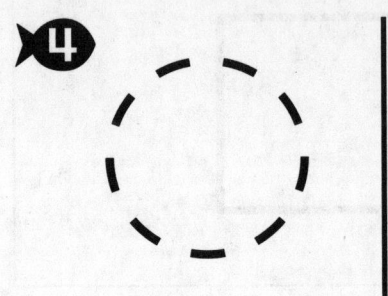

Math @ Home Activity

Draw a circle, hexagon, rectangle, square, or triangle on a sheet of paper. Have your child trace the shape and then draw three copies of the shape. Repeat with different shapes.

Directions: 3. How can you draw a square? Trace the square. Then draw a square. **4.** How can you draw a circle? Trace the circle. Then draw a circle.

Additional Practice

Name _____

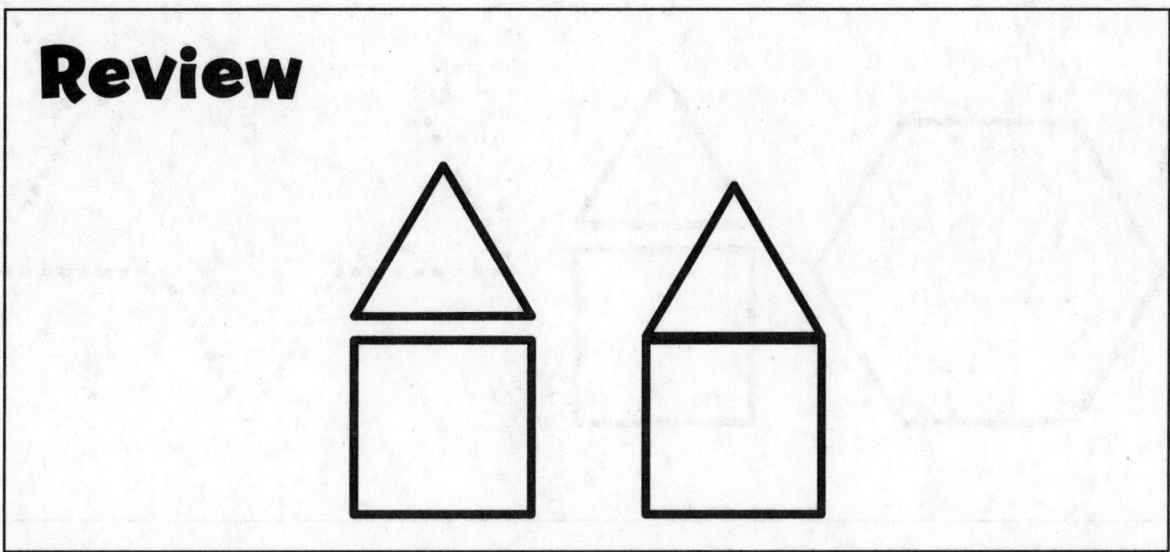

Review

❶

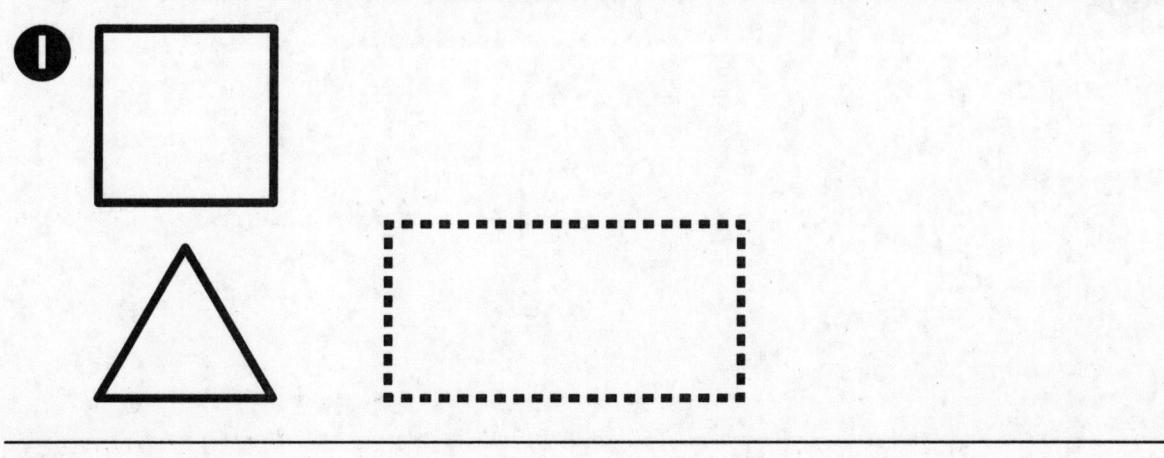

❷

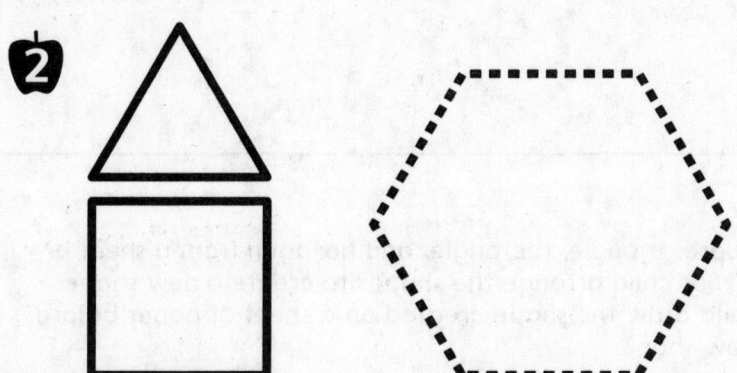

Review: You can use different types of shapes to make a new shape.

Directions: 1–2. How can you use pattern blocks to make a new shape? Choose one type of pattern block to make the new shape. Trace the pattern blocks you used.

3

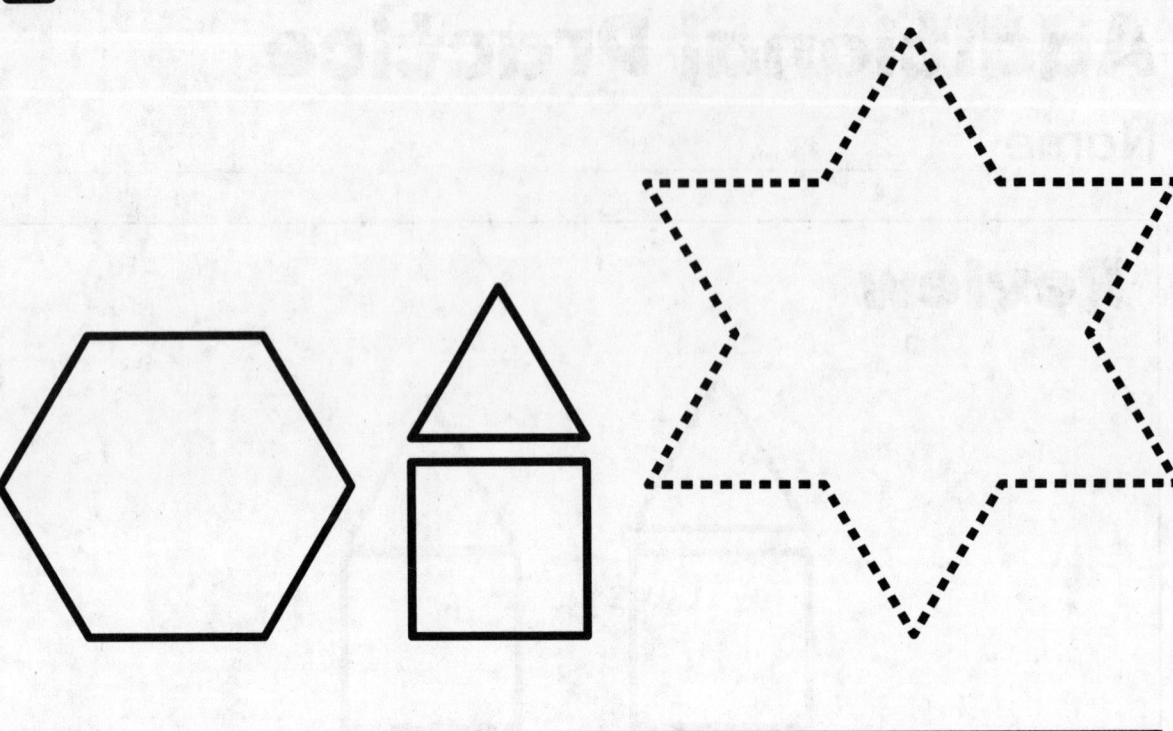

 4

 **Math @ Home Activity**

Cut out a square, triangle, rectangle, and hexagon from a sheet of paper. Have your child arrange the shapes to create a new shape. Have your child draw the shape created on a sheet of paper before creating a new shape.

Directions: 3. How can you use pattern blocks to make a new shape? Trace the pattern blocks you used. **4.** How can you use square pattern blocks to make a larger square? Trace the pattern blocks you used.

Additional Practice

Name _____

Review

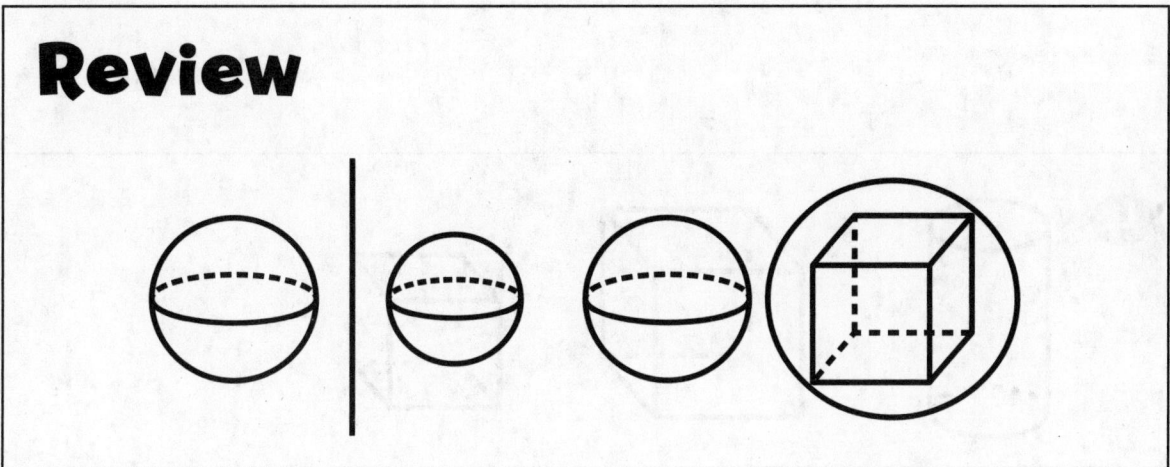

❶

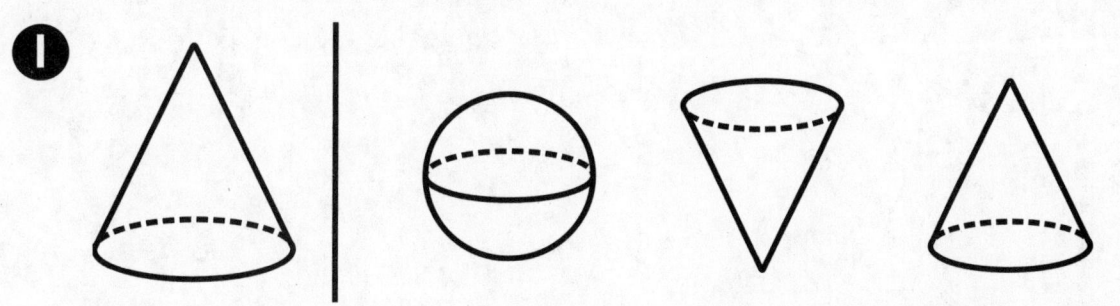

❷

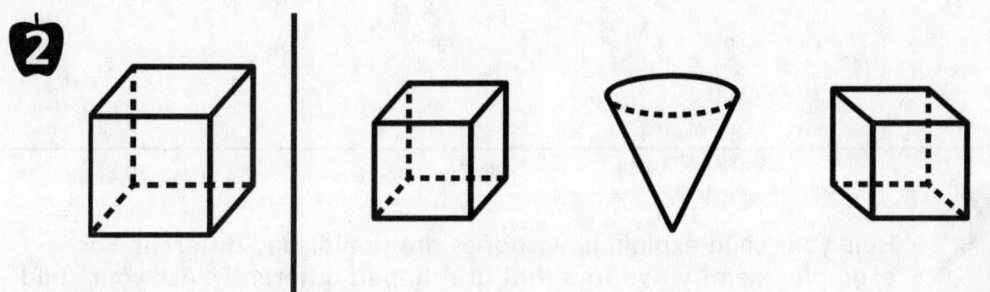

Review: You can compare 3-dimensional shapes. The shape that is circled is different from the first shape.

Directions: I. How can you compare the shapes? Draw an X on the shape that is different from the first shape. **2.** How can you compare the shapes? Circle the objects that are like the first shape.

3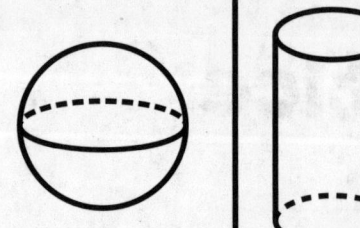

4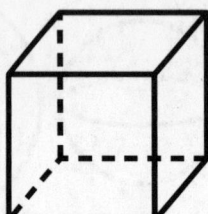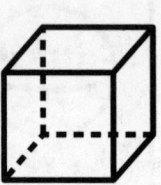

Math @ Home Activity

Help your child explain how shapes are similar and different. For example, identify two toys that are shaped differently. Ask your child to explain how they are different. Then identify two toys that are shaped similarly. Ask your child to explain how they are alike.

Directions: 3. How can you compare the shapes? Circle the shape that is like the first shape. **4.** Which shape is different? Draw an X on the shape that is different.

Additional Practice

Name ....

Review

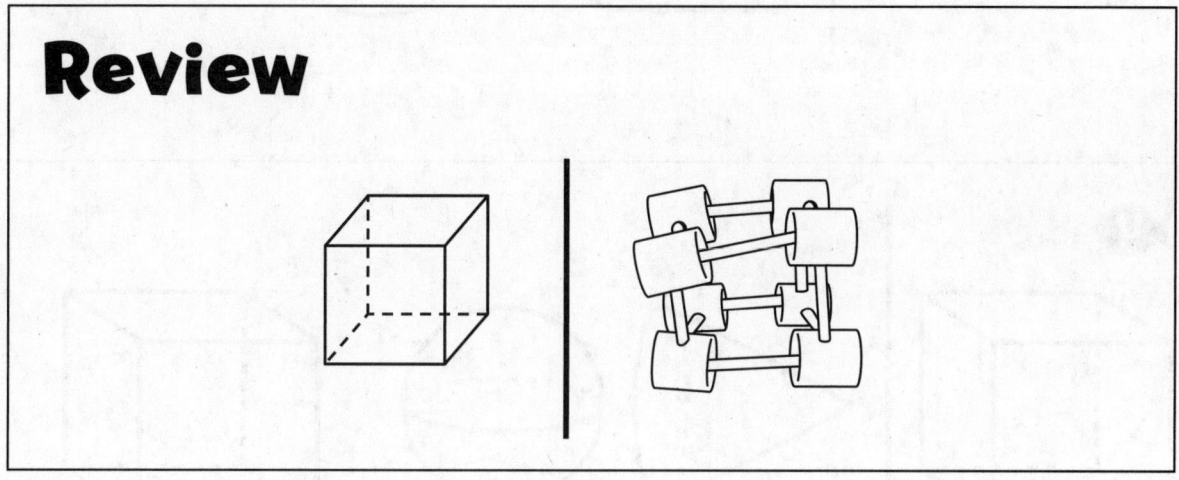

1

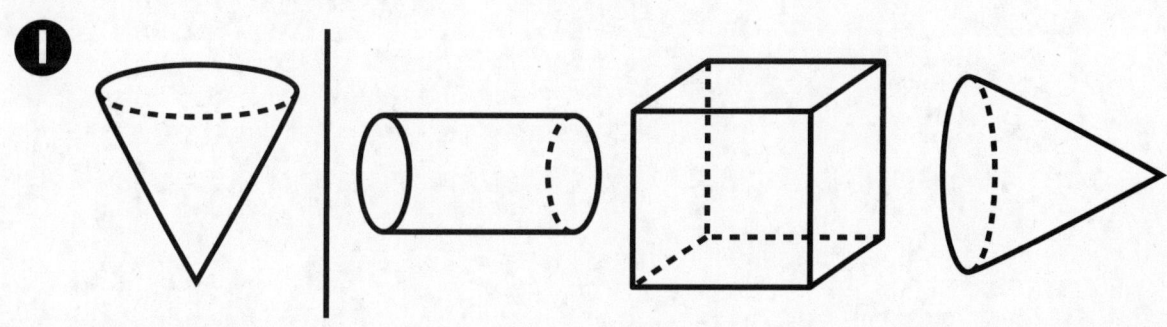

2

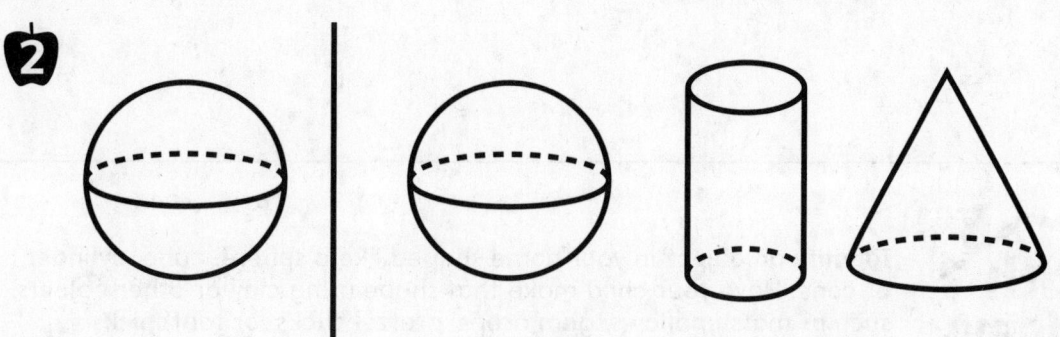

Review: You can use clay or other objects to build 3-dimensional shapes.

Directions: 1. How can you build a cone? Use clay to build a cone. Circle the solid shape you built. **2.** How can you build a sphere? Use clay to build a sphere. Circle the solid shape you built.

3

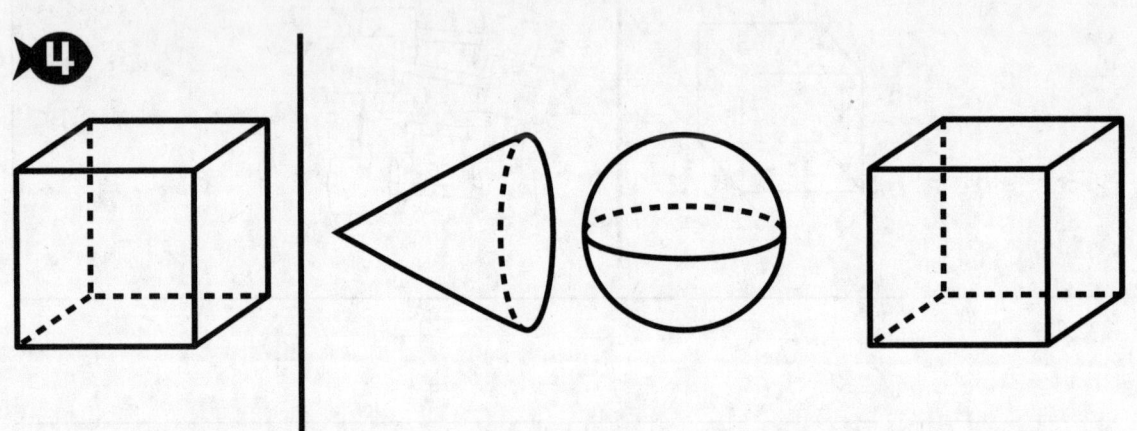

4

Math @ Home Activity

Identify an object in your home shaped like a sphere, cube, cylinder, or cone. Have your child make that shape using clay or other objects, such as marshmallows, gum drops, pretzel sticks, or toothpicks. Repeat the activity with a different shape.

Directions: 3. How can you build a cylinder? Use clay to build a cylinder. Circle the solid shape you built. **4.** How can you build a cube? Use clay to build a cube. Circle the solid shape you built.

Additional Practice

Name _____

Review

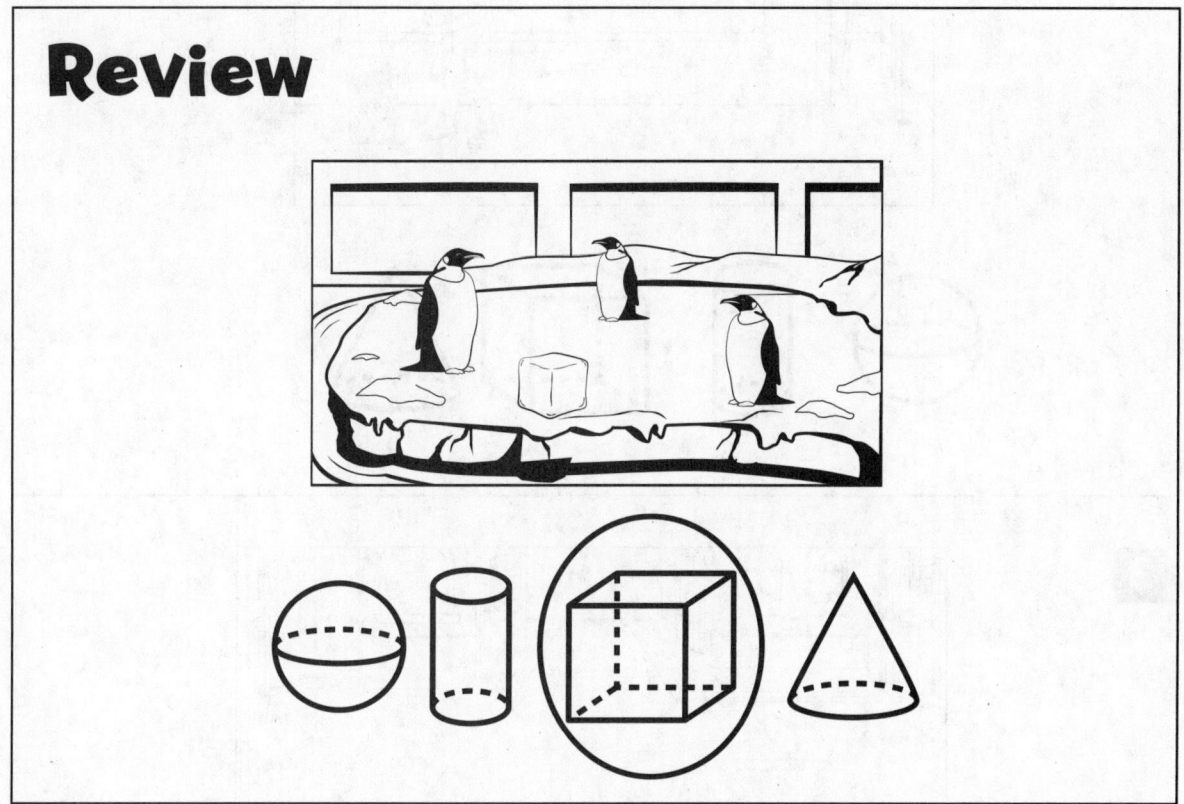

1

  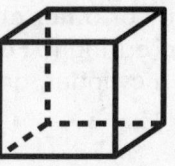 

Review: You can find 3-dimensional shapes in a picture. The block of ice looks like a cube.

Directions: 1. What shape does the baseball look like? Circle the shape. Name the shape.

2

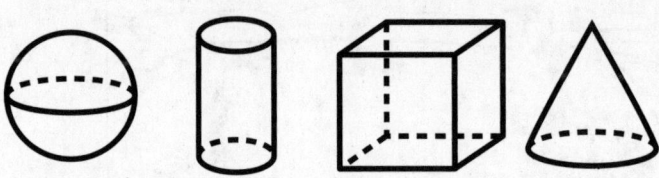

3

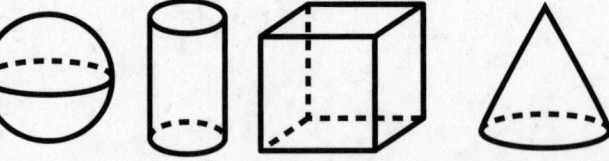

Math @ Home Activity

Have your child look around your house to identify a shape they can represent using clay or other objects. Have your child say the name of the shape before creating it. For example, your child might find a soup can, identify it as a cylinder, and then build it out of clay or salt dough.

Directions: 2. What shape does the wastebasket look like? Circle the shape. Name the shape. **3.** What shape do the traffic cones look like? Circle the shape. Name the shape.

Additional Practice

Name ....

Review

1

Review: You can identify objects you can measure by length, height, weight, and capacity. You can measure the length of a carrot. You can measure the height of an ear of corn. You can measure the weight of an apple. You can measure the capacity of a mug.

Directions: 1. Which objects can be filled to measure capacity? Circle the 2 objects that could measure capacity.

2

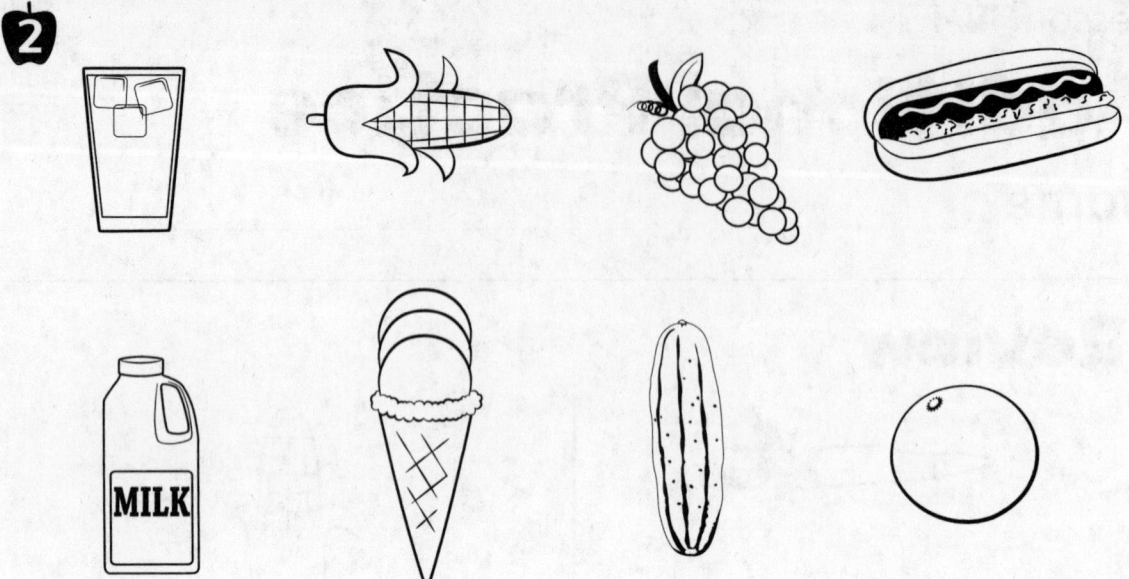

3

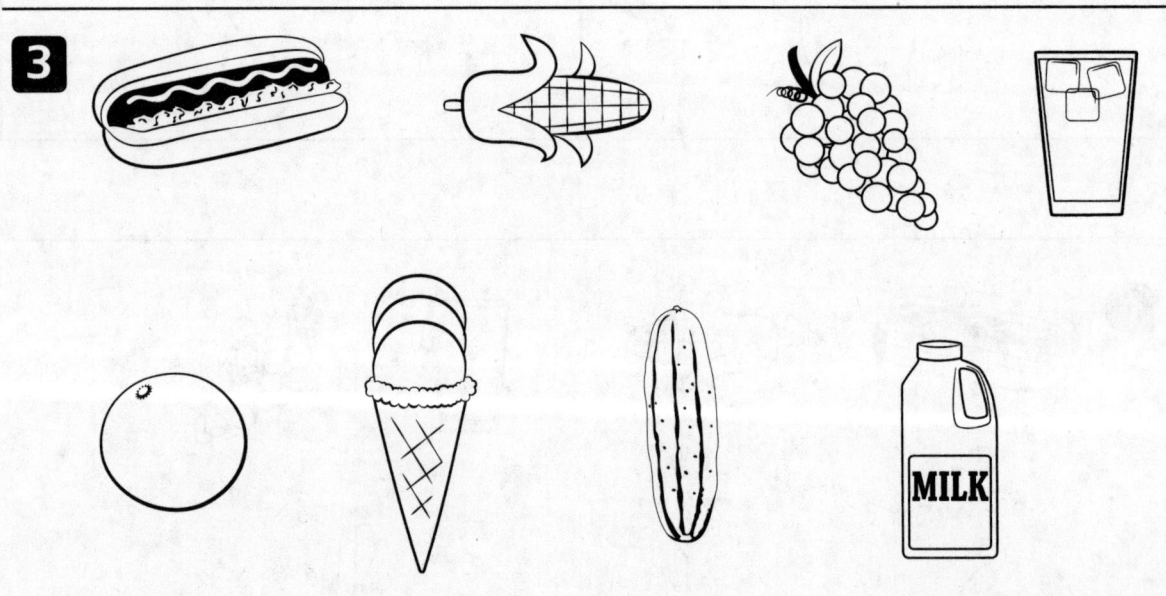

Math @ Home Activity

Identify an everyday object and ask your child to identify whether he or she would measure its length, height, weight, or capacity. Most objects can be measured in multiple ways. Encourage your child to explain his or her thinking.

Directions: 2. Which objects can be measured by length? Draw a square around two objects you can measure by length. **3.** Which objects can be measured by weight? Put an X on two objects you can measure by weight.

Additional Practice

Name _____

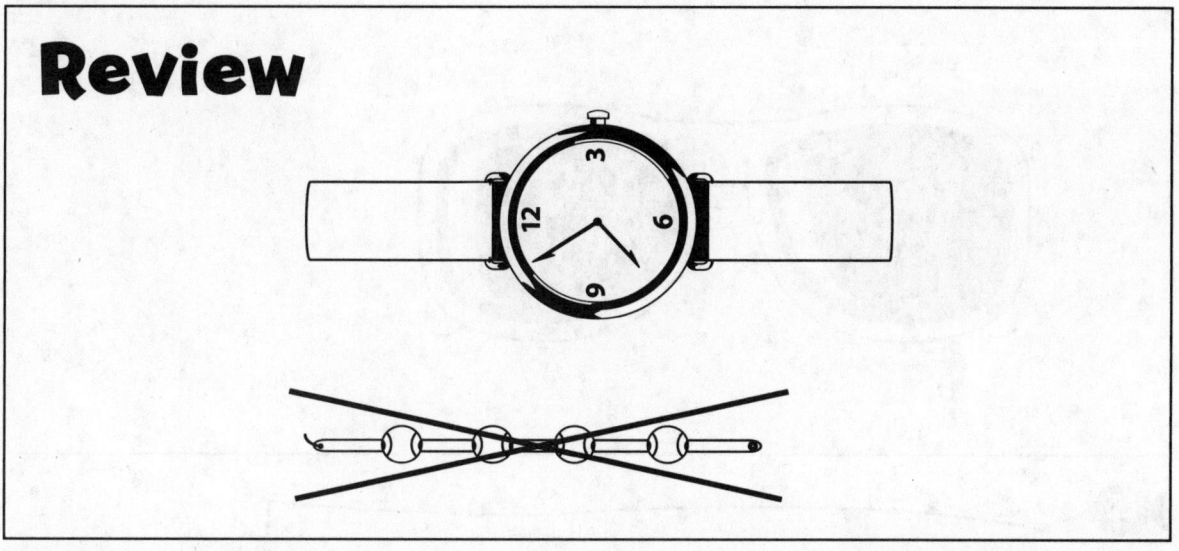

Review

---

①

②

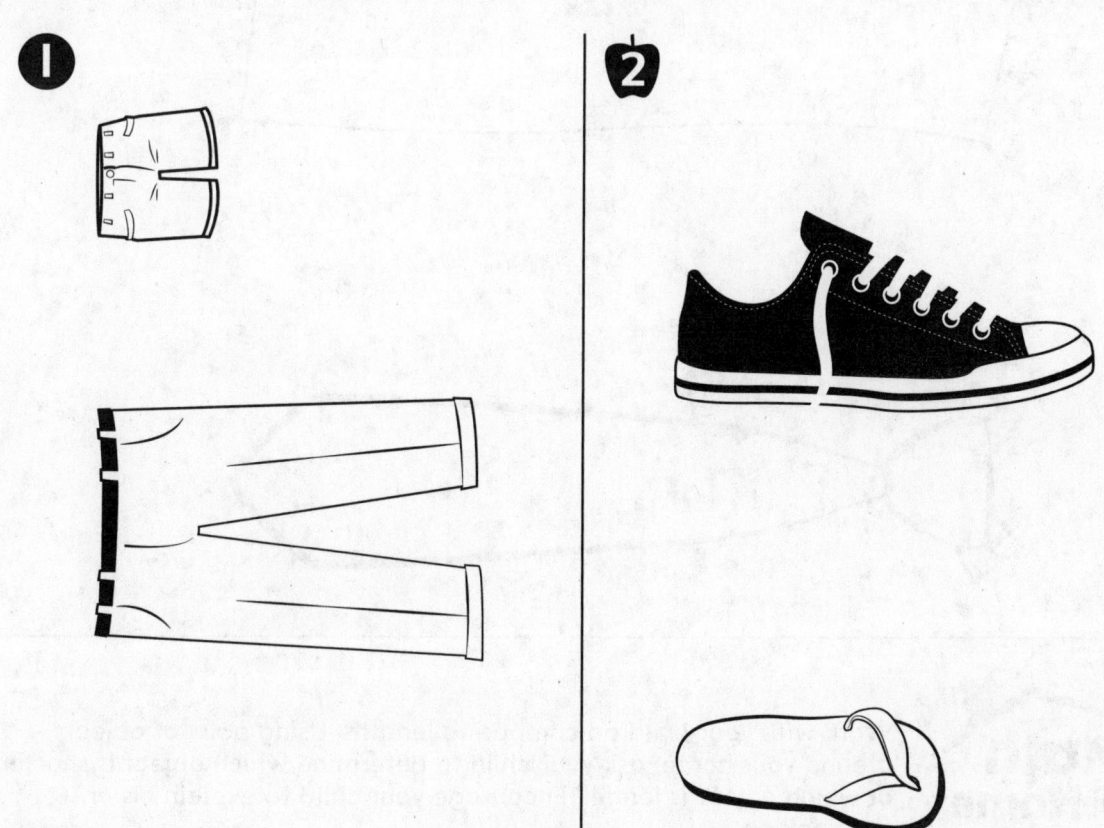

Review: You can compare the lengths of objects to identify which one is shorter.

Directions: 1–2. Which is longer? Circle the longer object. Underline the objects if they are the same length.

3

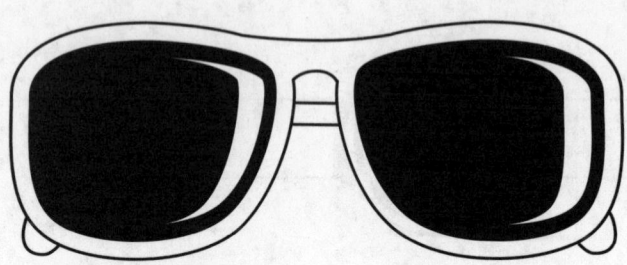

4

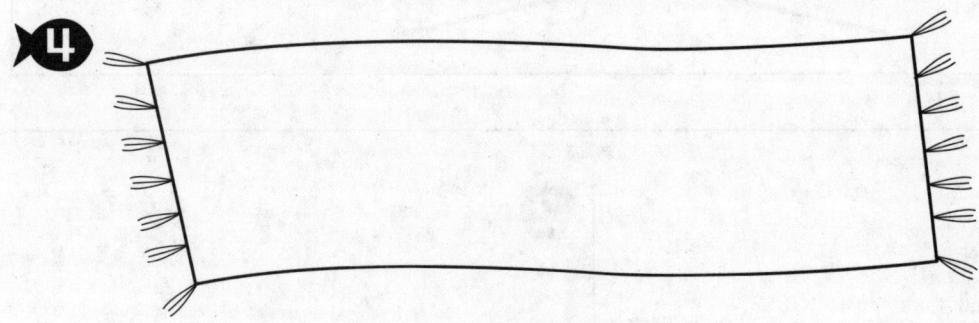

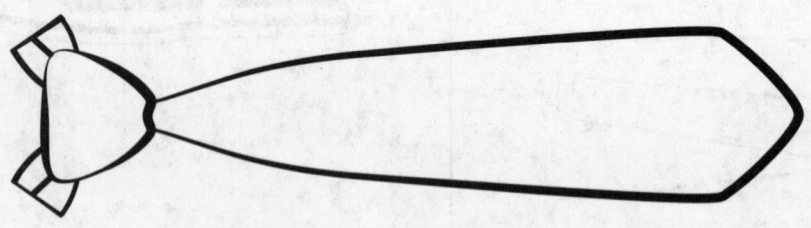

Math @ Home Activity

Work with your child on comparing lengths. Using pairs of objects around your home, ask your child to determine which object is shorter or which object is longer. Encourage your child to explain his or her thinking.

Directions: 3–4. Which is shorter? Draw an X on the shorter object. Underline the objects if they are the same length.

Additional Practice

Name _____

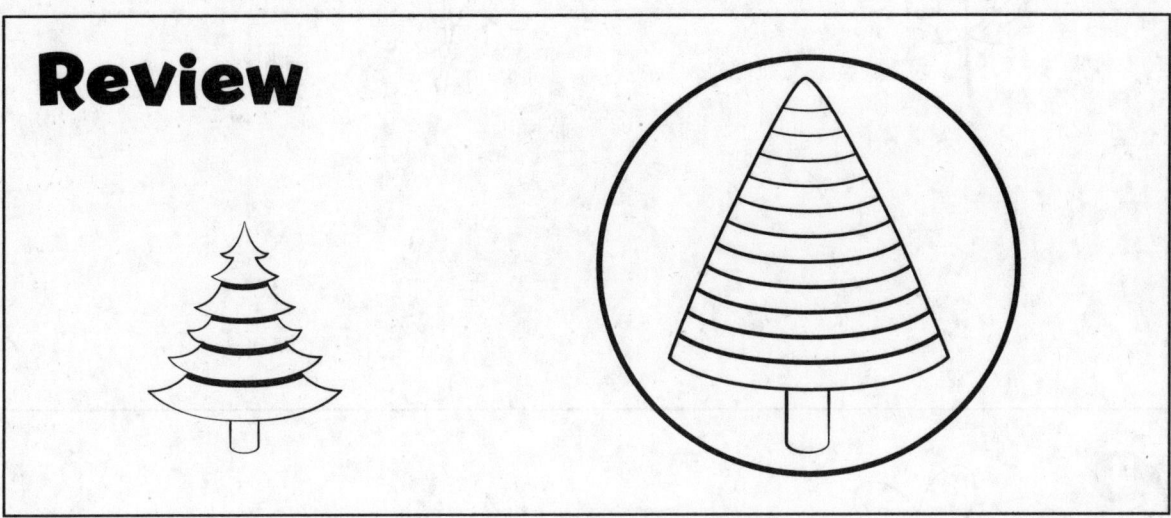

Review

1

2

Review: You can compare the heights of objects to identify which one is taller.

Directions: 1–2. Which is shorter? Draw an X on the object that is shorter. Underline the objects if they are the same height.

3

 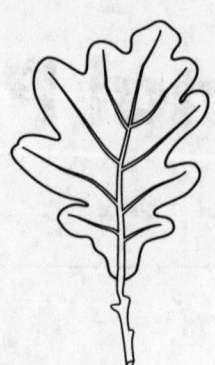

Copyright © McGraw-Hill Education

Math @ Home Activity

Draw two objects on a sheet of paper. Have your child determine which object is taller. Then have him or her draw two objects and ask you which object is shorter. Continue the activity, taking turns drawing.

Directions: 3–4. Which is taller? Circle the taller object. Underline the objects if they are the same height.

Additional Practice

Name _____

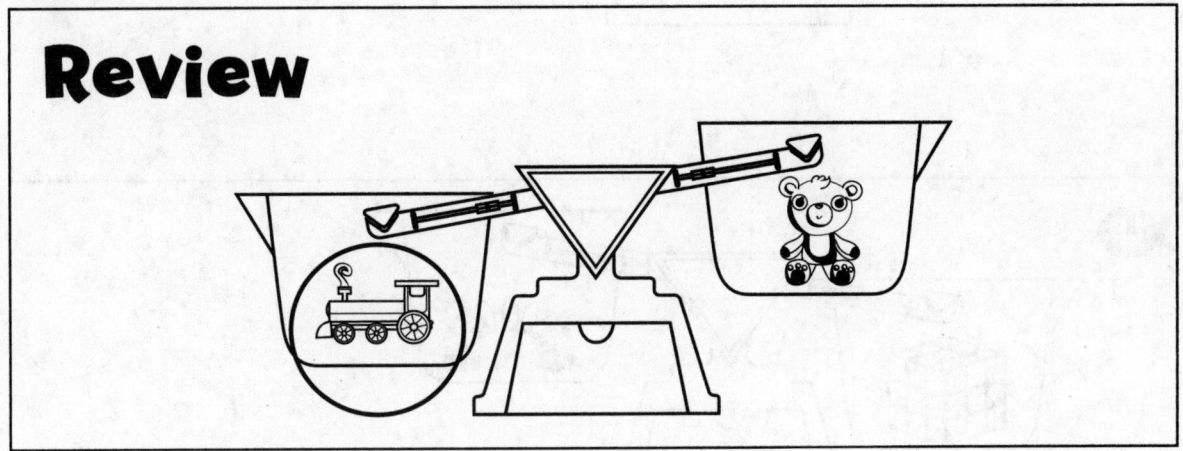

Review

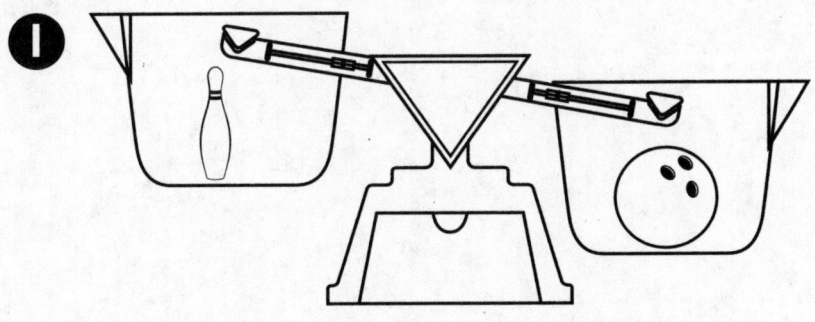

❶

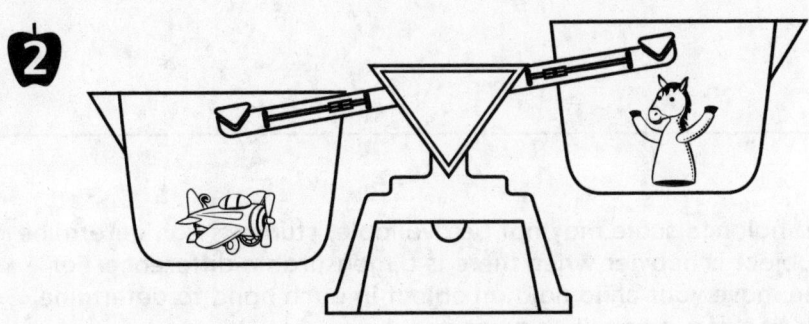

❷

Review: You can compare the weights of objects to identify which one is heavier.

Directions: 1. Which is lighter? Draw an X on the object that is lighter. Underline the objects if they are the same weight. **2.** Which is heavier? Circle the object that is heavier. Underline the objects if they are the same weight.

3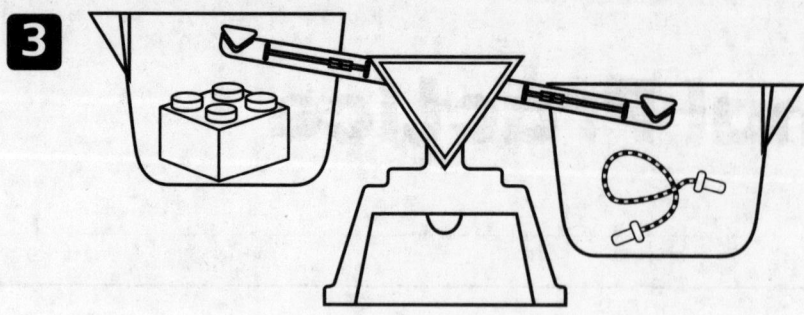

4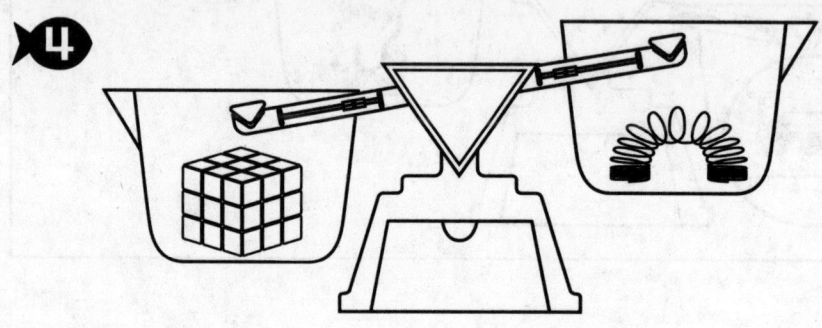

Directions: 3. Which is heavier? Circle the object that is heavier. Underline the objects if they are the same weight. **4.** Which is lighter? Draw an X on the object that is lighter. Underline the objects if they are the same weight.

Lesson 14-5

Additional Practice

Name

Review

1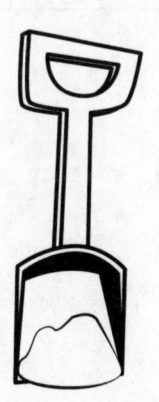

2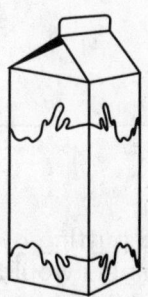

Review: You can compare the capacities of objects to identify which one holds less.

Directions: 1–2. Which holds more? Circle the object that holds more. Underline the objects if they have the same capacity.

3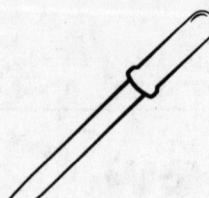

***

 4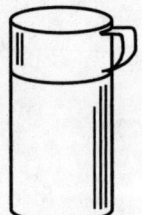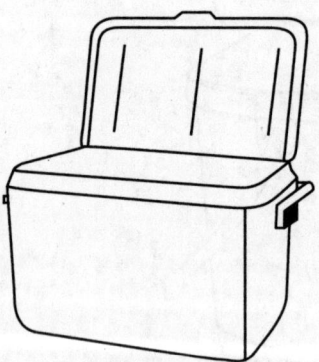

***

Directions: 3–4. Which holds less? Draw an X on the object that holds less. Underline the objects if they have the same capacity.